Anthropology, Ecology, and Anarchism
A Brian Morris Reader

Anthropology, Ecology, and Anarchism: A Brian Morris Reader
© 2014 PM Press.

ISBN: 978–1–60486–093–1
Library of Congress Control Number: 2013956919

Cover by John Yates / www.stealworks.com
Interior design by briandesign

10 9 8 7 6 5 4 3 2 1

PM Press
PO Box 23912
Oakland, CA 94623
www.pmpress.org

Printed in the USA by the Employee Owners of Thomson-Shore in Dexter, Michigan.
www.thomsonshore.com

To the memory of an inspiring teacher,
Mr. L.W. Bennett, who encouraged me to read books

Contents

Acknowledgments

I am appreciative of the journals and magazines that have published my various writings over the last decade, and convey my warm thanks to the editors of the *New Humanist*, *Critique of Anthropology*, *Democracy and Nature*, *Social Anarchism*, *Anarchy*, and *Anarchist Studies*. I should also like to thank my family and many friends for their support and encouragement, and to especially thank Angela Travis and Sue Lloyd for seeing the book through the press.

Introduction
by Peter Marshall

Brian Morris has done many things but he is above all a professor emeritus in anthropology. Before doing his doctorate research on hunter-gatherers in Southern India he was a tea planter in Malawi and this led to a lifelong interest in the people, animals, and insects of that country. He has had a passionate concern with natural history and ecology as well as anthropology. His many writings include the full-length studies *Forest Traders* (1982), *Anthropological Studies of Religion* (1987), *Western Conceptions of the Individual* (1991), and *Anthropology of the Self* (1994). He says that he came to anarchism in the 1960s when he met an anarchist who persuaded him of the importance of freedom. I first met him in the 1980s at the Anarchist Research Group at the Institute of Historical Research at London University and we have kept in touch on and off ever since. He has not only written about the natural historians and pioneers of what he calls "Ecological Humanism" but full-length studies of the Russian anarchist thinkers Bakunin and Kropotkin. He seems most inspired by the latter thinker, especially by his emphasis on mutual aid and his deep ecological sensibility. He has already written in 1996 a collection of essays called *Ecology and Anarchism*.

The present collection, combining his interest in anthropology, ecology, and anarchism, is therefore to be warmly welcomed. Many of the essays and reviews have been published in comparatively obscure journals and are therefore usefully brought together here. They are scholarly yet accessible; erudite yet the learning is worn lightly; historical yet often addressing burning issues of the day. As ever Morris is a prolific polymath, thoughtful, original, and full of insight. His style is not turgid and full of jargon like many academics in his field but flows naturally and reads well. He is informed, eclectic, and widely read.

These essays show a close relationship between anthropology, anarchism, and ecology (the human subject, society, and the natural world). As Morris makes clear, anarchism is not a utopian dream or terrorist nightmare; it is firmly based in the real world and shows the potential of all human beings to live in a more just, equal, and free society close to nature and all that live within it.

The first essay, analysing theorists in anthropology, psychology, and sociology, concerns the rise and fall of the human subject. He concludes that there is little to be gained by taking an extreme position either expunging the human subject from critical analysis or collapsing him or her in cultural relativism. Culture does not pre-exist the human subject but they dialectically coexist.

In the second essay Morris defends "realism" and "truth." It is an important work of methodology but it helps the reader to have some knowledge of the debate. Anthropology has never been a monolithic discipline. Yet postmodernism, which claims to have no knowledge of the world except through discussion and argues that there is nothing outside the text, has become very fashionable among some academics. It is against this rather nihilistic vision which suffers from historical amnesia that Morris writes. He wishes to affirm the importance of "realistic metaphysics" and claims there is such a thing as truth and representation, human agency and empirical knowledge. Our engagement with the world is always mediated; there is therefore no direct relationship between language and the world. He sees truth as a form of representation. This is not to confuse it with objectivism or a form of scientism. While we may affirm the independent existence of reality it does not mean we do not see it through our language and consciousness. All knowledge is an approximation of the truth. As Marx said long ago, "Men make their own history, but they do not make it just as they please." Human agency is therefore important but the individual is not a floating atom but an embodied being; there is a dialectical relationship between humans and the social structures they create.

What is needed, Morris argues, is therefore an approach that combines science (naturalism) and hermeneutics (humanism). Humans are an intrinsic part of nature, and their social life can be explained in terms of empirical science. Morris follows a long tradition that is concerned with the interpretive understanding and scientific explanation; in a sense he has tried to unite enlightenment and romantic traditions while rejecting crass textual realism and positivism. At the same time, he has a strong sense of history and tradition. While recognising the unity of humankind he does not deny our difference and singularity. He concludes that anthropology affirms a realist ontology and recognises that anthropological understanding is ultimately a search for truth.

In the third essay, "The Anthropology of Anarchism," he makes the familiar distinction made long ago by Harold Barclay between anarchy, which is an ordered society without government, and anarchism, which is a political movement which came to the fore in the last century. Morris rightly points out that there were several anarchists who were also anthropologists, such as Élie Reclus, brother of Elisée Reclus, and Célestin Bouglé who wrote a study of Pierre-Joseph Proudhon as a pioneer sociologist. Peter Kropotkin, Murray Bookchin, John Zerzan and others also drew on anthropology more or less successfully to support their arguments. Morris then offers some personal reflection on anarchism, pointing out that anarchism has traditionally been opposed to the state capital and the church. As Michael Bakunin wrote, "freedom without socialism is privilege and injustice; and socialism without freedom is slavery and brutality."

Morris in this collection is clearly a social anarchist or a libertarian socialist, which he considers synonymous. Anarchism of course has been misrepresented as destructive, violent, and nihilistic but is none of these things. Morris makes a useful distinction between what the French noun calls *puissance*, meaning power over others, and the verb *pouvoir*, meaning the power to do something. Anarchists are opposed to all coercive power in the first sense but not necessarily in the second sense as the ability to get things done. Anarchy is not therefore a synonym for chaos, as it is often portrayed by its opponents, but the most ordered form of society. As Proudhon wrote, "It is liberty that is the mother, not the daughter, of order." As for being violent it is no more so than other creeds and there has been a long pacifist tradition. And far from being anti-intellectual it offers a rich and complex system of ideas and beliefs and has attracted some of the finest intellectuals of the age. Most anarchists have also been feminists and more recently supported the green and peace movements. Although the rise of the nation-state was a recent development in history, anarchists like Gustav Landauer and Colin Ward have argued that anarchy without recourse to coercive authority has always been with us, like flower seeds ready to grow in a desert with life-giving rain.

In his essay on Buddhism, anarchy, and ecology, Morris offers critical reflections on Buddhism. He is right to draw a distinction between the teachings of Buddha and the behaviour of Buddhist monks who have aligned themselves with states, whether they be the Emperor Asoka, who built his empire on violence, or contemporary Buddhist monks and clergy in Sri Lanka and Myanmar (Burma), some of whom have been the worst nationalists. Buddha claimed no divine inspiration; strictly speaking, Buddhism is atheistic in recognising no God. Buddha hoped that his teachings would be discarded as a raft once one had become enlightened. He did however oppose desire, aversion, and ignorance, which can only bring suffering in this world, and taught the doctrine of no-self and encouraged an ascetic life of frugality and peace. Above all, Buddhism has generally been a process philosophy and had a dialectical approach. While Buddha advocated detachment from this world he did not believe that it is an illusion as in Hinduism. Buddhists traditionally have not engaged in agriculture and expected others to do the dirty work.

By his own admission, Morris does not paint a very "rosy picture" of Buddhism, but he recognises that there are aspects of the teachings of Buddha which are anarchistic. Buddha rejected, for instance, state structures and recommended seeking one's own salvation without texts or teachers. His view of nonviolence, his compassion for all living beings and his stress on generosity, frugality, and peace of mind still have contemporary relevance.

In another essay, Morris offers a tribute to "three pioneers of ecology," the sociologist Lewis Mumford, the biophysicist René Dubos, and Bookchin, whom he calls social ecologists and "down-to-earth." They all accepted Darwin's evolutionary theory and believed that human beings are a product of natural evolution but claimed there is a cultural evolution as well. They remained true to the tradition of the Enlightenment and were committed to free enquiry, reason, and science. As such they believed in universal ethics and the unity of humankind. In a sense both Mumford and Dubos were religious thinkers but not Bookchin; the former were essentially radical liberals while the latter was a social ecologist. But while they recognised that humans are an integral part of nature and the relationship between humans and nature was symbiotic, they still talked of the uniqueness of the human species. Mumford and Dubos thought that we should be stewards of nature, and Bookchin wanted to "steer" the course of evolution.

In essay eight Morris discusses "people without government," a phrase made popular amongst anarchist circles by the anthropologist Harold Barclay. While maintaining that all societies have some form of power, Morris rejects all coercive forms of power which are based on violence and found in societies with centralised government, hierarchy, and domination. He refutes the notion that humans are naturally good, which we find in Jean-Jacques Rousseau, as well as the view that humans are selfish, competitive, and aggressive, as in Thomas Hobbes. In contrast, Morris discusses three ethnographic contexts that exemplify people without government: foraging communities, village-based societies that are focused on horticulture, and chiefdoms. He concludes his essay by discussing eco-feminists' accounts of "matriliny" and the Mother-Goddess religion associated with the earth. We do not entirely agree here, as my research for *Europe's Lost Civilization: Uncovering the Mysteries of the Megaliths* suggests that the Mother-Goddess religion was widely held by the early peoples bordering the Mediterranean and during the late Neolithic period in Northwest Europe when people first settled down to agriculture.

In his review of Joel Kovel's book *The Enemy of Nature* Morris points out that to call capitalism the enemy of nature is hardly new or original. It has a good account of ecological crisis which is largely the results of globalisation, but the author fails to understand the nature of anarchism which has always been anti-capitalist as well as anti-state. Kovel's plea for a new brand of Marxism is to make it more ecological but he adopts a form of market socialism. As Gustav Landauer said, "Marxism is a professor who wants to rule."

Another readable review is of Simon Tormey's *Anti-capitalism: A Beginner's Guide* which is said to provide a very helpful discussion of the movement in all its diversity. But looking at the anti-capitalist movement, Morris points out that the majority of protesters are not anti-capitalist. They are the reformist rather than revolutionary and want to make capitalism more benign. Most want to increase the economic and political power of the nation-state. Tormey not only devotes a mere seven pages to anarchism but makes the ridiculous claim that Marx was a "true anarchist."

In his reflections on the so-called "new anarchism" or "postmodern anarchism," Morris argues that there is in fact precious little which is original and which cannot be found in the traditional class-war and social anarchist thinkers. Rejecting those who have been influenced by postmodernism and post-structuralism, Morris claims that the new anarchism is hardly a paradigm shift, except perhaps in their exponents' lack of historical awareness, their moral relativism, their theoretical incoherence, and their rejection of universal truths and values. Even the Situationists, with their emphasis on *détournement* and poetic resistance find their antecedents in the French Revolution. Although Friedrich Nietzsche influenced the individualism of Emma Goldman, and he attacked the nation-state, he nevertheless was committed to the aristocratic elite. Morris is particularly critical, as an anthropologist himself, of Zerzan's so-called eco-anarchism in which he calls for a return to the life of the hunter-gatherer before the foundation of agriculture. Zerzan, according to Morris, is not only historically inaccurate but the present population of seven billion humans can hardly turn to that way of life in the future. As Morris says, "post-anarchism is simply an exercise of putting old wine into the new wine bottles!"

Amongst the anarchist thinkers Morris clearly prefers the "gentle anarchist" Rudolf Rocker. From 1895 Rocker was an anarchist missionary to the Jews in the East End of London, editing the Yiddish anarchist newspaper *Arbeter Fraynd* (Worker's Friend). During the First World War he was imprisoned as an enemy alien but after the war returned to Germany where he became a leading exponent of anarcho-syndicalism. With the rise of Nazism Rocker went to America where he published his great work

Nationalism and Culture, which Bertrand Russell thought was an important contribution to political philosophy.

Rocker recognised that anarchist ideas had existed throughout human history but modern anarchism as a political movement emerged in the nineteenth century. It was for him a confluence of socialism and liberalism. On the one hand, socialism wanted to put an end to economic monopolies; on the other hand, liberalism celebrated the freedom of the individual. Anarchism, which Rocker considered a synonym of "libertarian socialism," could therefore only be achieved in a society based on voluntary association and mutual aid. He dismissed organised religion—"all power has its roots in God"—but emphasised the need for religious tolerance; indeed, the "spiritual nature" of humans expresses itself in culture, that is to say, in science, art, literature and every branch of philosophical thought and aesthetic feeling. Above all, Rocker is remembered for his work on nationalism, which he thought was "the religion of the democratic state." As an anarcho-syndicalist, he emphasised class struggle and the need for trade unions not only to defend the rights of the producers but to prepare for the reconstruction of social life. In later life Rocker tended to replace anarcho-syndicalism with what he called "community socialism."

Morris further looks at the political legacy of Bookchin in another essay. He sees his early work as seminal in developing social ecology by linking the degradation of the natural world with the exploitation of humans. One cannot dominate human beings without dominating nature. Morris recognises that Bookchin however became a controversial figure in later life, particularly in his stress on "libertarian municipalism" and his demand that his followers participate in local elections and be bound by majority vote. Bookchin reached his lowest point in his 1995 essay "Social Anarchism or Lifestyle Anarchism: An Unbridgeable Chasm." He labelled "lifestyle anarchists" as egoistic liberals. He finally even rejected the description of being an anarchist.

In his essay "Ecology and Socialism," Morris explores the relationship between the two. He rejects however the interpretation of an "ecological crisis" in terms of religious metaphysics, which seems to him as a new rendering of the concept of original sin. Again, Morris argues against the deep ecologists as well as my book *Riding the Wind: A New Philosophy for a New Era*, which he sees as a reaffirmation of ancient religious traditions. "Liberation ecology," as I call my new philosophy, is of course much more than this, and I give a very anarchistic account of social and economic arrangements of a free, just, and ecological society.

Instead, Morris prefers Darwin's "evolutionary ecology," which depicts humans as an intrinsic part of nature, and the "revolutionary socialism" of Kropotkin and the "social ecology" of Bookchin, who linked the ecological crisis with the capitalist system. The answer in his view is anarchism with an ecological perspective and which stresses like Bakunin that socialism without liberty only leads to tyranny and state capitalism.

At the same time, Morris criticises my history of anarchism *Demanding the Impossible* as applying the term "anarchist" to a wide variety of different philosophies and individuals. In essay fourteen Morris says that "anarchism is fundamentally a historical movement and political tradition that emerged only around 1870." He appears to define very narrowly anarchism as applying to "anarchist communists"; indeed he regards as synonyms the terms "anarchism, anarchist communism, and libertarian or revolutionary socialism." Although he is not always consistent, Morris would seem to reject here as "anarchists" the philosopher William Godwin, the individualists Max Stirner and Benjamin Tucker, the mutualist Pierre-Joseph Proudhon, the first to call himself self-consciously an anarchist, the collectivist Michael Bakunin, and the pacifist Leo Tolstoy because they were not communist. Yet they are all part of the anarchist tradition, which cannot simply be reduced to "anarchist communism" or "revolutionary socialism." As Morris recognises elsewhere, an "anarchist sensibility" has existed from the earliest times in the great river of anarchy as Kropotkin, Rocker, and George Woodcock among others have recognised.

Finally, Morris is on firm ground in his memories and reflections on the foragers of South India, particularly the Malaipantaram, which was the basis of his PhD fieldwork and his book *Forest Traders*. He claims that they had an individualist ethic, not in a non-social sense but in the stress they put on the autonomy of the individual as an independent person. At the same time, a sense of equality permeated their society. Indeed, the economic exchanges within their society may best be described in terms of reciprocity and mutual aid. They had a marked antipathy towards any formal authority or hierarchy whether based on wealth or power. They may have appealed to the ancestral spirits of forest deities for support and protection in times of illness or lack of food but they generally had a very practical approach to the natural world. They could be called communistic in their sharing ethos and wish not to exploit others or natural resources. Their stress on personal autonomy of the individual did not exclude mutual aid and social solidarity. Morris concludes that individualism, egalitarianism, and communism are manifested in the social life of the Malaipantaram South Indian foragers and that they undoubtedly lived in a state of anarchy, in a sense of a society without government, hierarchy, and domination, and practised voluntary cooperation and mutual aid. These are clearly values which Morris appreciates.

As the above introduction makes clear, I may not always have the same emphasis as Morris. His humanist ecology emphasises the importance of human beings whereas my liberation ecology has a more spiritual dimension in the sense that I am concerned not only with our place on earth but also in the wider universe. I see no great contradiction between social and deep ecology. I believe there is no "unbridgeable chasm" between so-called "lifestyle anarchism" and "social anarchism"; indeed, one usually leads to the other. Social anarchism need not be equated with "class struggle" anarchism, as Morris asserts, since one can believe that class war is not as relevant today in Western societies as it was before. Tired of the unnecessary nit-picking and sectarian squabbles, I prefer to adopt a form of "anarchism without adjectives" as advocated by Voltairine de Cleyre, Errico Malatesta, and others. I can tolerate the coexistence of many different anarchist schools.

Yet it is right that Morris and I should sometimes disagree. He appears however in this collection to be a dialectical thinker and recognises that anarchism thrives best on discussion and opposition as well as agreement and mutual aid. What Morris has to say about anarchism, anthropology, and ecology in this collection is well worth reading, discussing, and acting upon. It will undoubtedly clarify one's thinking and inform many debates about its subject matter. It is therefore very timely and to be warmly welcomed.

Preface

Throughout my life I have had three essential interests—intellectual interests, that is.

The first is an absorbing interest in ecology, or rather in natural history, and I have been exploring and enjoying the natural landscape and its wildlife since I was a boy. One of the first books I ever possessed at the age of fourteen was *Look and Find Out: Birds* by W. Percival Westell. And significantly the first article I ever published was in the field of natural history. It was published in a wildlife conservation magazine, *African Wildlife* (1962), and was titled "Denizen of the Evergreen Forest." It described the life-history and ecology of a rather rare pouched mouse, *Beamys hindei*.

My second interest has been anthropology, for I have always been fascinated by the diversity of human culture and have always had an especial interest in philosophy and religion. But I have long been intrigued by such mundane topics as wild foods, animal traps, totemic signs, how to make fire without matches, medicinal plants, and particularly the culture of the Native American Indian. I was especially inspired as a teenager by Ernest Thompson Seton's *The Book of Woodcraft and Indian Lore* (1912). Books such as *Forest Traders* (1982), *Western Conceptions of the Individual* (1991), and *Religion and Anthropology* (2006) are the result of my anthropological interests.

My third interest has been anarchism, which I discovered at the age of twenty-nine, when, quite by chance, I met Bill Gate. It was at a meeting on comprehensive education at Conway Hall in London. Bill was an affable working class bloke, rather large and Bakunin-like, and he introduced me to what for me then was a rather esoteric political tradition, namely anarchism. I have been an anarchist (or libertarian socialist) ever since, and from the 1960s have been actively involved in many protests and demonstrations, whether against the Vietnam War or the poll tax. But essentially I have tried to live my life as an anarchist.

All my writings over the past forty years have in one way or another been interdisciplinary as I have sought to bring these three interests of mine together. Thus, my ethnographic researches in Malawi have focused largely on natural history topics—medicinal plants, fungi, mammals, and insects—and I have sought to relate these to the social and cultural life of the Malawian people. When nearly two decades ago I published a collection of my various essays and reviews—originally published largely in obscure radical magazines and journals—it was significant that I should title it *Ecology and Anarchism* and dedicate it to Bill Gate.

Also important to note is that in all my writings I have tried to bridge the gap between academic scholarship and a wider lay public (whether students or radical activists) and have therefore always tried to write in a way that was lucid and accessible as well as scholarly. I have thus never taken the advice given to me long ago by one anonymous academic referee that I should write in a style in order to impress academic scholars.

What I find significant is that the scholars and writers who deeply influenced me during my youthful years were also involved in bringing together my own interests and were thus, in various ways, pioneer ecologists. These scholars were the literary naturalist Richard Jefferies, the artist-naturalist Ernest Thompson Seton, and the anarchist geographer Peter Kropotkin. I have published studies of all three radical ecologists.

The present book brings together some of the essays and reviews that I have published over the last decade in various radical magazines and academic journals, along with some earlier work. All initially were given as talks to either anarchist groups or at meetings of various socialist organisations. In a way it brings together my three interests—anthropology, ecology, and anarchism—and thus offers critical reflections on a range of topics and contemporary issues. As with my earlier collection, I trust all my essays and reviews are charitable and fair, and will be of interest to scholars, students, and radical activists alike.

1

The Rise and Fall of the Human Subject (1985)

Is there in the human sciences a specific psychological approach or level of analysis? Is there a conceptual space for the human subject?[1] There are two extreme positions on this issue. One gives the abstract individual, or subject, or personality variables, priority as factors in explaining human behaviour or culture (see Kaplan and Manners 1979, 127–59). This strategy is assumed by existentialists, ego psychologists, and some psychoanalytic writers, personality structures being most commonly employed as intervening variables. It is also implicit in much transactional analysis—such as the early writings of Barth and Leach, where the *abstract* individual is seen as an agency of social change. Often described as methodological individualism, this position has serious flaws.[2]

The other extreme position expunges the subject from the analysis entirely, and sees sociocultural phenomena as an autonomous realm, completely independent of psychological variables. This is the viewpoint of behaviourism and sociobiology, which put great emphasis on the biological analogy of "organism and environment" and not only play down cultural variables but find little space for the concept of person or even the conscious mind. This is why there has been a strong reaction against behaviourism by existentialists and humanistic psychologists. It is also the viewpoint of many anthropologists within the Durkheimian tradition, for Durkheim saw culture as a phenomenon *sui generis*, and in reacting against any kind of psychological reductionism eliminated—in theory— psychological variables from the analysis of culture. "Man is double. There are two beings in him: an individual being which has its foundation in the organism . . . and a social being which represents the highest reality in the intellectual and moral order" (1915, 16). Social facts could therefore only be explained by other social facts, and the social realm is seen as independent of both psychological and environmental factors. In countering the suggestion that this therefore left social life "in the air," Durkheim drew an analogy between this social/individual duality (which he stresses) and the relationship between mental faculties and the brain, arguing that we cannot reduce the "mind to the cell and deny mental life all specificity" (1974, 26–29). But he could hardly affirm—if he followed his own reasoning and analogy—that there is no relationship between these two entities or variables.[3]

The anthropologist who has taken this kind of cultural emphasis to the extreme is Leslie White, for whom the individual is no more than—as the Gestalt psychologist Köhler (1937) put it—"an empty container" for the products of the group. White (1949) distinguishes essentially between three distinct levels of reality, the physical, the biological, and the cultural—and psychology is firmly associated with the second category, as one of the biological sciences, or as a kind of analysis that links the individual organism with cultural facts. The individual as such is but "the expression of a cultural tradition in somatic form," and human behaviour simply the response of the human organism to "extra-somatic, symbolic stimuli which we call culture" (1949, 139). The human individual is therefore neither the creator nor a determinant of culture; "he is merely a catalyst and a vehicle of expression." Like Durkheim, White stresses that culture is an entity *sui generis*, with "a life of its own," and more than once he writes that to study culture scientifically one must proceed "as if the human race did not exist" (1949, 209).

For all these writers there is a radical separation of the individual from the cultural setting. The theoretical orientation is essentially positivistic (see Allport 1955, 7–12). Yet some recent developments within the Marxist tradition have stressed a viewpoint similar to that of White and the behaviourists. Whereas Marxists in the Hegelian or socialist humanist tradition—Sartre, Marcuse, and Fromm—have advocated a dialectical approach to social life and have viewed history as the project of the human subject, some recent Marxists, following Althusser, have viewed the concept of the "human" as essentially a part of bourgeois ideology and hence unscientific. History, for Althusser is a "process without a subject" (1972, 183), and he is equally keen to reject any notion of human nature in general, or at least any notion which implies that the individual has any explanatory role in the science of history (see Callinicos 1976, 66–71; Thompson 1978, 193–406). This anti-humanism has gone hand-in-hand with a rejection of historical time or process; it is the advocacy of a static structuralism within a highly mechanistic discourse.

Many writers (e.g., Coward and Ellis 1977) have uncritically accepted Althusser's formulations, presenting familiar ideas as if they had only just been discovered by structural Marxism. The notion that the subject or person is a social construct, and that in different societies or at different periods of European history there have been varying conceptions of the human individual, is hardly new. The category of "human essence," or of the abstract individual, that transcends any given social context, which Coward and Ellis see as a fundamental premise of idealism was, of course, associated not with idealism but with early bourgeois philosophy of a materialistic kind. Criticisms of this notion are not a recent innovation of structuralism but are pre-Marxist. Stirner, a precursor of atheistic existentialism, made some salutary criticisms of this kind of liberal humanism more than a century ago (1845, 123–29). Marx, in his criticisms of Feuerbach, simply replicates Stirner's reservations about abstract humanity, and in *The German Ideology*, Marx and Engels subjected Stirner himself to the same searching critique. But rather than seeing the notion of the human subject as a "metaphysical fiction" (Hirst and Woolley 1982, 131) that must be expelled from scientific discourse, Marx sought a more genuine humanism, stressing that the essence of man involved the discharge of a creative practice that was essentially social (see Novack 1973; Fromm 1970, 68–84). Behaviourism, Durkheimian sociology, sociobiology, structural Marxism and Lévi-Straussian structuralism thus have much in common in their repudiation of the human subject. (Although for Lévi-Strauss the concept of the objective "mind" retains a crucial role [see Shalvey 1979, 35].) Coward and Ellis express this kind of anti-humanist perspective rather well, arguing, like Durkheim, that society is prior to the individual, and that the human subject is "constructed." Rather than being a critique of bourgeois idealism, this simply replicates it, for culture is only prior to the individual if culture is conceived abstractly as uncreated by humans, and the individual is conceived as a *particular* asocial organism. But the human subject is an "emergent" entity that makes culture possible. Culture does not "pre-exist" the human *subject*; they dialectically coexist.[4] As Piaget insisted, an "epistemic subject" is essential for a meaningful analysis of social life: "if cognitive structures were static, the subject would indeed be a superfluous entity" (1971, 70). The alternative is either a collapse into cultural relativism (and indigenous psychologies), or the positing of a transcendental subject—which is what Coward and Ellis evidently have in mind in their critique of the subject. But there is I feel a way between the static ahistoric structuralism they seem to espouse with subjectivity expunged—and the kind of phenomenology that gives the human subject absolute priority.

What is of interest about these two theoretical treatments of the human subject is that they represent, in extreme form, two tendencies within the human sciences—the humanistic and scientific—and that they show that the social sciences also exemplify Gellner's "pendulum swing" theory. If we examine the history of the social sciences over the past fifty or more years it appears to have moved through four distinct phases. In the first phase, around the turn of the century, anthropology and psychology happily coexisted as disciplines; there was a salient interest in human consciousness (evident particularly in the writings of William James and Boas); and many of the founders of experimental psychology—Rivers and Wundt—also undertook anthropological studies, without confounding the two approaches. In the second phase, however, after the First World War—when psychology, anthropology, and sociology were establishing themselves as independent academic disciplines, and psychoanalysis and Marxism were finding an institutional footing—the human "subject" was hardly mentioned. Structural-functionalism became the dominant paradigm in both sociology and anthropology—in spite of the influence of Malinowski—and there developed a kind of phobia about psychology (Lewis 1977, 2). Psychology itself was seemingly entrenched in Watson's style of behaviourism in which concepts such as "consciousness" and "mind" were deemed to denote the unobservable and therefore to be inadmissible in scientific discourse. Marxism too, under the influence of the mechanistic materialism of Engels and Plekhanov, saw little scope for a science of the psyche.

During the 1930s a changing emphasis can be discerned in all the human sciences, and attempts were made in various ways to bring the human "subject" back into the analysis of culture. I shall discuss this third phase in some detail, focusing on the writings of four scholars—Kardiner, Reich, Fromm, Laing—and only briefly mention the latest phase, namely that associated with the dominance of structuralism, which has rejected the historical subject entirely.[5]

Abram Kardiner

During the 1920s anthropologists and psychoanalysts were natural allies in the intellectual revolt against the constraints of sexual and other forms of provincialism. Boasian anthropologists "enjoyed a reputation for Bohemianism which they earned as proponents of the relativity of morals, as taboo-breaking feminists, or as practitioners of exotic custom" (Harris 1969, 431). Benedict, Mead, and Herskovits, in particular, were strong advocates of cultural relativism. Two things happened. One was that there was a continuing move away from evolutionary or materialist analyses towards a more humanistic approach. "Civilisation" then became, in the work of Freud and Benedict, a synonym for culture, rather than a term that depicted Western society or capitalism (as it had been for Morgan and Engels). Secondly, Freudian theories and concepts became an "irresistible lure" and came to permeate more and more anthropological studies. Psychoanalytic theory was a "major stimulus" to anthropology (Hallowell 1976, 212) and to the social sciences generally. By the 1940s some kind of rapport between psychoanalysis and social science (and Marxism) had occurred, and it took various forms.

Within anthropology Freudian theory had a major influence on the "culture and personality" school. This school includes those who did not explicitly or systematically accept Freudian doctrines and concepts, but whose writings are nonetheless permeated with his ideas. The notions that childhood experiences to a large extent determine the adult personality, and that religion, folklore, and witchcraft are "projective" systems, were the two key ideas which were imbibed by the anthropological tradition. Kluckhohn's study of *Navaho Witchcraft* (1944) for example is infused with Freudian themes, and Margaret Mead's later writings, particularly *Balinese Character* (1942), written with Bateson, are "saturated with psychoanalytic terms, concepts and nuances" (Harris 1969, 434). But the writer who came to represent most clearly this kind of approach, in which psychoanalytic concepts became "embedded," as it were, in the analysis of social life, thus giving rise to a kind of *psychosocial* interpretation, is Abram Kardiner. Trained as a psychoanalyst, Kardiner was instrumental in the late 1930s in organising a series of seminars on "culture-and-personality" in conjunction with a group of anthropologists, of whom Ralph Linton and Cora du Bois were among the most prominent. The idea behind the seminar was that the anthropologists of the group should present ethnographic data on the cultures they were familiar with, which would then be "analysed" by Kardiner.

Although a psychoanalyst, Kardiner seems, in developing his theory, to have abandoned all the basic tenets of psychoanalysis: the Oedipus complex and the libido theory in particular are not mentioned at all. For Kardiner the most signal and durable aspect of Freud's theory was the idea of approaching human life from the "point of view of biography" and in establishing criteria for the "study of the character of the individual" (1945, 11). Given the anthropological framework and the concept of culture Kardiner inherited from Boas, it is not surprising that the central idea of his psychodynamic analysis of culture was the "basic personality structure." This is defined as "the effective adaptive tools of the individual which are common to every individual in the society" (1939, 237). It thus refers to those personality characteristics that are shared by a majority of members of a particular community and which are the result of formative childhood experiences. The concept, Kardiner admits, is but a refinement of an old idea, going back at least to Herodotus, and generally known as "national character." This concept, Kardiner suggests, introduces a relativistic factor into the conception of history, enabling us to do away with the idea of a uniform and constant "human nature" (1945, 415). It also has a constraining effect, he suggests, on the type of adaptation that a community will make in specific historical circumstances. This basic personality is essentially formed in the earliest years of life: it is the creation of what Kardiner termed the "primary" institutions. These included such aspects as "family organization, in-group formation, basic discipline, feeding, weaning, institutionalized care or neglect of children's anal training, sexual taboos . . . subsistence techniques" (1939, 471). The Freudian influence is apparent here, and "primary institutions" are focused less on the economic infrastructures than on child-rearing practices.

There is some truth in Voget's suggestion that Kardiner was seeking some kind of "middle ground" between Róheim's extreme ontogenetic approach, in which virtually *all* aspects of culture were interpreted in terms of childhood experiences, and that of Benedict who saw culture as moulding the person into its mirror image (Voget 1975, 441). As Kardiner himself put it: "the individual stands *midway* between institutions which mould and direct his adaptation to the world, and his biological needs, which press for gratification" (1939, 17). The primary institutions, which are instrumental in shaping the basic personality structure, and which have their "focus" within the individual personality, are taken by Kardiner as given. He offers no explanation or interpretation to account for these primary institutions and, a bit like Róheim's ontogenetic analysis, it is a theory without a "ground floor." Kardiner himself admits that only history can throw light on the forms which these institutions took, and that he knows of no satisfactory theory to account for them (1939, 471). But Kardiner does attempt to provide an explanation for the secondary institutions—religion, folklore, mythology, art—for these are seen as "projections" fashioned and structured by the basic personality. Thus culture is split into two aspects: the primary institutions which determine the basic personality or structure, and the secondary institutions, which are expressive of this personality as a "projective" system. LeVine (1973, 55–58) calls this the "personality mediation" view. Harris applauds, and sees promise in Kardiner's attempt to bring religion and ideological practices within a deterministic framework, and concludes his critique of Kardiner with a plea to unite psychodynamic theory and historical determinism (1969, 442). This is the intended project of Fromm and the early Reich, whom Harris does not mention.

Kardiner, in trying to tie "the concept of history to a psychological base" (1945, 415), and in viewing the human personality as a "mediating" variable between two kinds of institutions, attempted to steer his analysis between two extremes. On the other hand he was critical of the functionalism of

Malinowski, Kroeber, and Benedict, and also of White's notion that cultural phenomena interacted—"without reference to the human species"—an idea he thought ridiculous.[6] On the other hand he is equally critical of Róheim's ambitious attempt to employ Freud's libido theory to explain social phenomena, on the grounds that this strategy leads to confusion and faulty interpretations. Neo-Freudian analyses, he felt, did not lead anywhere, and Fromm's theoretical ideas were simply a mixture of Martin Buber and Zen Buddhism (Kardiner and Preble 1961, 213).

As a theoretical movement within anthropology the "culture and personality" school hardly survived the crisis of the 1950s when it came under critical attack from several directions (Bock 1980, 141). Although Devereux has suggested that a revival of the culture and personality approach is urgently needed in the "overly intellectualistic climate of current anthropology" (1978, 394) it may rightly be regarded as a particular phase in the development of anthropology. I have discussed Abram Kardiner's work at some length because although his work is situated within the anthropological tradition, his style of analysis has close affinities to that of Reich and Fromm, both of whom attempted to introduce a psychological dimension to Marxism.

Wilhelm Reich

Described by Robinson as "one of the most volatile imaginations of the twentieth century" (1970, 19), and by Wilson as a "wayward and paranoid genius" (1981, 6), Reich was suddenly rediscovered in the late 1960s. His writings on "sexual politics" (he coined the term) then began to be translated and republished, and like Marcuse and Laing he became one of the intellectual doyens of the counter-culture movement, as well as an important influence on feminist writers (see Mitchell 1975, 137–226).[7]

Settling in Vienna in 1920, where he met Freud, Reich became a member of the Vienna Psychoanalytic Society and quickly established his reputation as a brilliant therapist as well as publishing articles on sexuality and therapeutic techniques in psychoanalytic journals. Being alive to the social and political realities of the period, with its mass unemployment and poverty, the social suppression of sexuality, and the repression of political protest, Reich began to get involved in politics. He established sex clinics where sexual advice and free psychoanalysis could be obtained, joined the Austrian communist party, and, through his writings and work, attempted to promote a "sexual revolution." In practice he attempted to "politicise" psychoanalysis, while stressing to the socialist movement the political importance of sexual reforms, and the understanding of psychological issues. He visited Russia in 1929 and made himself familiar with the writings of Marx and Engels, as well as dipping into the anthropological literature, particularly the studies of Malinowski, who became a firm friend and supporter of his work. The outcome was that between 1927 and 1933 Reich published five important studies, as well as a seminal paper on "Dialectical Materialism and Psychoanalysis" (1929). They represent a creative and enduring effort to draw together the insights of Marx and Freud into a meaningful synthesis. In his later years, after settling in the United States, Reich's writings took on a more pantheistic tone for he became convinced that a special kind of energy, *orgone*, was the primordial source of life—an idea similar to Bergson's *élan vital*, except that this energy was visible and measurable. His later writings have been described as "fanatical and bizarre" (Mitchell 1975, 152). Throughout his life, Reich was a romantic naturalist who attempted to steer between the extremes of materialism and mysticism.

In his early writings on sexuality, Freud had suggested that neurosis is the result of a conflict between instinctual demands and opposing official demands and implied that such conflicts were largely of a sexual nature. Reich adopted wholeheartedly this stress on the importance of sexuality in the understanding of neurosis, but he made three crucial amendments to Freud's libido theory. First he narrowed the definition of sexuality, stressing only the all-importance of *genital* sex. Second, he reversed Freud's procedure and emphasis, making actual and not psycho-neuroses the central focus of his own theory. And finally, Reich interpreted not only neurosis but all types of psychic disturbances in terms of sexual malfunctioning. He linked such malfunctioning, moreover, to "undischarged sexual energy." Psychic health therefore was seen as a direct outcome of what Reich termed orgastic potency—"the capacity for surrender to the flow of biological energy without inhibition, the capacity for complete discharge of all dammed-up sexual excitation." Mental illness is thus the result of a disturbance in the natural capacity for love, and "the core of happiness in life is sexual happiness" (1942, 114). It is not surprising therefore that Reich should interpret the Freudian unconscious largely in terms of instinctual strivings of a sexual nature, and view its effect as wholly positive. Given such premises Reich came to present a theory which is virtually the antithesis of Freud's.[8]

In *The Mass Psychology of Fascism* (1933) Reich is concerned not only to understand fascism, but also to raise a more general question: what socioeconomic function did sexual repression serve? His answer was that the repression of sexuality was not, as Freud contended, for the creation of "culture" in general, but rather "forms the mass-psychological basis for a certain culture, namely the patriarchal authoritarian one, in all its forms" (1930, 10). Fascism, he believed, should not be discussed as a lot of mystic nonsense. Nor could its hold on the majority of working people in Germany be explained in terms of Hitler's demagogic powers, the machinations of big business, or the political blunders of liberal politicians. Socialist theory he felt to be quite inadequate to understand why broad elements of the working class had embraced fascism. Marxism was clearly at fault in that it "had failed to take into account the character structure of the masses and the social effect of mysticism" (1933, 39). He therefore turned to Marx's concept of ideology and to Freudian theory—or rather to his own interpretation of it—for theoretical support.

His basic argument was that ideology—"the mainstay of the state apparatus" (1933, 59)—had a material force, and a locus in the psychic structure of individuals. The family, which Reich considered to be the product of a particular economic system, has an essentially political function. Through sexual repression, it is a "factory" for authoritarian ideologies. The lower middle class German family in particular was especially accessible to nationalism and authoritarian ideology because of economic amenities (1933, 75). Following Malinowski, Reich came to argue that the Oedipus complex was not universal but specific to such an authoritarian culture. He saw religion and mysticism as reinforcing obedience to authority and as negating or substituting for healthy, affirmative sexuality. In short, the repression of sexuality within the patriarchal family, itself the product of definite economic circumstances, not only results in neuroses but also "makes people lack independence, will power and critical faculties" (1930, 78). This kind of character structure in turn supports the political and economic structure as a whole—authoritarian patriarchy. As with Kardiner, the family and the character-structure are seen as mediating between the economic structures and the ideological superstructures—in this case fascism.

Many writers have applauded Reich's imaginative and suggestive insights, which suggest that one should look at the family structure "to understand how economic realities are translated into religion, ethics and politics" (Robinson 1970, 43). Mitchell, too, suggests that Reich's critique of a patriarchal family has an "enduring value" (1975, 179). It is a theme later taken up by Laing and his associates. But Reich never really explains why fascism arose specifically at a particular time and place. Indeed it is difficult to assess Reich's analysis, because he had such an incredibly gross and simplistic conception of human history. Following the implications of Morgan's and Engels's theory of the family and drawing on aspects of Malinowski's ethnographic account of the Trobriand Islanders, Reich posits an original matriarchy. In this kind of society there is orgastic potency, and no repression of the genital love of children and adolescents, and consequently no neurosis, or sexual "perversions" (1932, 31–58). It is a primeval work-democracy. Mitchell claims that his analysis of the ethnography is confused and contradictory (1975, 177), but Reich himself argues that in this society one finds *both* kinds of family organisation and culture, and that the marriage exchange found in the Trobriands was the central economic mechanism in the social transformation to a patriarchal system (1932, 106). This, for Reich, is the great watershed in human history, not the transition from feudalism to capitalism. Since the supposed eclipse of matriarchal society the remainder of human history—the last six thousand years—has been characterised by authoritarian patriarchy. How "fascism" fits into this wide panorama is never specified.[9] In many respects Reich's historical account has the same status as Ovid's and Chuang Tzu's depiction of a past golden age before the rise of state systems. One pleads for some cultural relativism.

Reich saw the need to draw together Marxism and psychoanalysis in the analysis of fascism. He wanted especially to politicise psychoanalysis, while stressing to the Marxists the necessity of studying psychological factors. For both, this implied putting the family and sexual repression on the agenda, as factors in the understanding of authoritarian (capitalist) culture and ideology. In his paper "Dialectical Materialism and Psychoanalysis" he argued that these two approaches were not incompatible. (Some contemporary Marxists are still not convinced, although they accept his analysis of the function of ideology [Brown 1974, 80–95].) Reich argues that psychoanalysis is not simply an idealist product of a decaying bourgeois culture; rather he insists that psychoanalysis is in essence a radical psychology—unless "watered down" by the "ego"-psychologists such as Rank, Adler, and Jung who had abandoned the libido theory—which has materialist roots and a dialectical method like Marxism. Moreover he suggests that Freud's interpretation of the irrational and Marx's concept of ideology have much in common, for there is no question "in Marx of the material reality of psychological activity being denied" (1929, 19). It is a misunderstanding of Marx to interpret him as a mechanical materialist, and Reich quotes extracts from his theses on Feuerbach to support this. He does not however suggest a "synthesis" of Marxism and psychoanalysis: for the former treats social and the latter psychological phenomena. But "the two can act mutually as auxiliary sciences to one another" (1929, 15).

> Economy without active emotional human structure is inconceivable: so is human feeling, thinking and acting without economic basis. One-sided neglect of one or the other leads to *psychologism* ("only the psychic human forces make history") as well as *economism* ("only technical development makes history") (1930, xxvii).

Whether Mitchell and others are correct to suggest that Reich simply made an "amalgamation" of the two approaches or reduced both to the "sociology of the family" (1975, 215) is an open question. But Reich's

criticisms of Róheim show clearly that he was against any kind of psychological reductionism; and he was equally opposed to the kind of vulgar
Marxism that saw a simple and direct relationship between economic
structures and human consciousness and ideology. In a sense, psychoanalytic variables and his focus on the family mediate between the economic
infrastructure and cultural forms.

Erich Fromm

Erich Fromm expressed very similar views to Reich in denying the opposition between the individual and society. Neither psychological nor sociological approaches to human life were, he believed, adequate in themselves.

> The thesis that psychology only deals with the individual while soci-
> ology only deals with "society" is false. For just as psychology always
> deals with a socialised individual so sociology always deals with a
> group of individuals whose psychic structure and mechanisms must
> be taken into account (1970, 155–56).

Fromm tried to avoid and counter two extremes. One was the approach
of behaviourist psychologists and sociologists such as Durkheim who
"explicitly wish to eliminate psychological problems from sociology" and
thus "neglect the role of the human factor as one of the dynamic elements
in the social process" (1942, 11). Man, he wrote, "is not a blank sheet of
paper on which culture can write its text" (1949, 23). Fromm was therefore
against any kind of cultural relativism, but he was equally against the idea
that the nature of man was "innate" and based on "eternal forces rooted"
in human biology. There is no "fixed human nature." Granted that there
are physiological imperatives such as hunger and thirst, and the need
for sleep, which can be subsumed under the instinct of self-preservation,
there is still no inherent human nature as such. "Human nature is neither
a biologically fixed and innate sum total of drives nor is it a lifeless shadow
of cultural patterns to which it adapts itself smoothly: it is the product of
human evolution." Human nature is therefore a product of history. "But
man is not only made by history—history is made by man" (1942, 10–17).[10]

Thus Fromm, interpreting Marx, tried to suggest a theoretical formula
that was neither "ahistoric," suggesting a universalism that implies an
unchanging essence of man, nor relativistic, implying that the human
personality is but a reflex of social conditions (1962, 29). Aware that the idea
of "human nature" or the "essence of man" had fallen into disrepute and
had been challenged not only by psychologists and cultural anthropologists such as Benedict but also by Marxists, Fromm felt that contemporary
thought had "lost the experience of humanity" which underlay the radical
humanist tradition. These critics do not deny the unity of the human race,

he remarked, but "they leave hardly any content and substance to this concept of humanity" (1962, 28).

In all his writings Fromm explicitly argued for, and detailed, his conception of "human nature," both as the basis of his own ethical and political judgements, and as a basis for his social theory. He followed a humanistic interpretation of Marx, relying heavily on Marx's early writings on alienation (1970, 68–74), but he came to portray an "essence" of man which seems to me as tenuously connected with Marx's concept of "praxis" as it is with Freud's biologism. His approach is neither Marxist nor Freudian but a kind of radical humanism. In *The Fear of Freedom* (1942) he argues that purely economic or purely psychologistic interpretations of fascism are both untenable. What is needed is a theory that combines both.

> Nazism is a psychological problem, but the psychological factors themselves have to be understood as being moulded by socio-economic factors; Nazism is an economic and political problem, but the hold it has over a whole people has to be understood on psychological grounds (1942, 80).

Character structure and ideology were the two aspects of the problem from this psychological angle. His theoretical framework therefore implied three variables—economic, psychological, and ideological.

> We have assumed that ideologies and culture in general are rooted in the social character; that the social character itself is moulded by the mode of existence of a given society; and that in their turn the dominant character traits become productive forces shaping the social process (1942, 252).

Expressed again in other terms, psychoanalysis provides the missing link between the ideological superstructure and the socio-economic base (Jay 1973, 92). Such a formulation is very close to the perspective of Reich and Kardiner in seeing "personality" as a mediating variable.

R.D. Laing

At the end of the Second World War the "behaviouristic hiatus" in psychology came to an end, and there emerged a broad current of thought bearing a renewed interest not only in cognition and consciousness, but also in the self and the human subject. Although this changing emphasis in psychology was influenced by the incorporation of psychoanalysis into academia (in the course of which psychoanalysis itself became an ego psychology), the primary factor was the rise of existentialism.

Although it has its precursors in the writings of Stirner, Nietzsche, and Kierkegaard, existentialism only came into prominence at the end of the

Second World War. It was an expression of the intellectual malaise of those who reflected on the inexplicable horrors that had occurred in the thirty or so years that had just passed. Two world wars, the rise of fascism and totalitarian states, the slaughter of colonial peoples, the concentration camps and the extermination of six million Jews, Hiroshima and Nagasaki, all of which composed the fabric of contemporary history, had, in some quarters, given rise to a mood of extreme pessimism (Novack 1966, 3–17). Inspired by the phenomenological philosophy of Jaspers, Husserl, and Heidegger (which in anthropological terms stressed a focus on meaning and a return to *verstehende* or interpretative sociology), existentialism articulated a particular brand of social understanding. There was a stress on existence rather than essence, and existence was essentially portrayed as tragic, fragmented, meaningless, irrational. But along with the introspection and the feelings of alienation that the tradition conveyed, came a stress on the individual and on descriptive psychology, and a hostility to all forms of thought which appeared to imply that human life was determined.[11] Approaches from psychoanalysis and sociology (including Marxism) came in for much criticism, although thinkers of both disciplines had attempted to overcome the subjective/objective dualism. Nevertheless, the stress on "subjectivism" and the concept of the individual as ontologically free to express his being-in-the-world (to use the kind of phrasing beloved by phenomenologists) introduced a humanistic dimension to social thought. The early writings of Jean-Paul Sartre formed a central focus of this tradition and provided a stimulus for new orientations in many intellectual fields. This humanistic tendency is noticeable in the writings of Fromm and in the Hegelian Marxists, and had a decisive influence in psychology, giving rise to a current of thought known as humanistic psychology. The work of Rollo May, Carl Rogers, and Abraham Maslow are particularly important in this respect.

R.D. Laing's study *The Divided Self* exemplifies this changing emphasis within psychology. He attempts to lay the foundations for a "science of persons," employing an existentialist-phenomenological perspective. For this purpose he draws explicitly on the work of Sartre and implicitly on the writings of H.S. Sullivan and his "interpersonal theory of psychiatry." Laing advocates a humanistic psychology. This involves the study of a person as a person—and the acceptance that "psychology is the logos of experience." From the start then we have a fundamental ambiguity. On the one hand we have a focal emphasis on the human person; a theory which "begins and ends with the person." On the other hand we have a stress on experience, on interpersonal relationships. Thus the "self," which is the key concept for Laing, is both a given entity and the product of social interaction.

> I wish to define a person in a twofold way: in terms of experience,
> as a centre of orientation of the objective universe; and in terms of
> behaviour, as the origin of actions (1967, 20).

For a full understanding of the human condition, I do not think there is a way out of this contradiction. Several writers have drawn attention to, and criticised, the "metaphysics of individualism" that they see lurking in Laing's work. Hirst and Woolley for example argue that Laing is mistaken in viewing the "self" as a directive entity, the inherent origin of action and the reference point of experience. There is, they write "no inherent conscious locus 'behind' conduct and independent of social personality which is their real origin" (1982, 205). Mitchell (1975, 256) is also critical of Laing's assumption that there is an "essential self," a discrete entity pre-existing all forms of communication. Just as it is misleading to conflate empirical knowledge with empiricism, so it is equally unhelpful, I think, to collapse the "human" subject into individualism.

> If there is no inherent individual to be realized once the shackles
> of existing social organization are thrown off then human beings'
> capacities and welfare become a matter of what forms of social
> organization should exist or be developed (Hirst and Woolley 1982,
> 205).

To avoid the collapse into cultural and moral relativism, and to provide some leverage for a critical appraisal of contemporary institutions, a humanistic perspective is both necessary and difficult to avoid.

Laing was not in any simple sense a "romantic individualist," having only a conception of the individual as an asocial being, though his social theory, like Fromm's, eventually degenerates into religious mysticism and moral aphorisms. But what is significant about Laing, for my present purpose, is that, like several other humanistic psychologists (and such sociologists as Irving Goffman), he attempted to put the "self" into the centre of social analysis without losing sight of the fact that the human subject was a social product.

By the 1960s the human subject had thus been re-instated solidly within the human sciences, and the "individual," "personality," or self was accorded a crucial theoretical role in several disciplines. The Neo-Freudian analyses of Kardiner and Fromm, the transactional analyses of Barth and Leach, the radical writings of Wilhelm Reich, and the humanistic interpretations of writers such as Laing, Bettelheim, Erikson, and Goffman expressed a similar perspective in their emphasis on human subjectivity. A reaction against the excesses of this kind of humanism inevitably occurred; two key figures in this reaction are Lévi-Strauss and Althusser, both of whom renounced the "subjectivity" and the historical dimension of the Hegelian tradition that Marx had inherited. Thus we can return now to the issue broached at the beginning of this article: is it possible to situate within the theoretical discourse of the human sciences concepts such as subject, mind, self, and psyche which imply a psychological level of analysis? In order to answer this question one must avoid the scientistic approach of structuralists in which the human subject is expunged from social analysis entirely, and the extreme humanistic approach—whether this is expressed in "subjectivist" or "individualistic" terms. Marx, in his advocacy of a humanistic science, suggested an approach that avoided these two extremes.

An extreme "humanistic" approach was suggested by early bourgeois social theorists such as Hobbes and empiricists such as Feuerbach. Various natural attributes were postulated as being the "essence" of humankind. This approach was basically asociological, and implied a dichotomy between the individual and society which Durkheim later bequeathed to his successors.[12] Marx criticised early bourgeois philosophy not for its idealism—in fact he applauded the materialism of Hobbes, Bacon, and Feuerbach—but for universalising what were specific historical circumstances and beliefs, and for ignoring the fact that what was fundamental about humans was neither god nor biology, but their social attributes. Thus in *The German Ideology*, Marx and Engels, having defined the productive forces as "a historically created relation of individuals to nature and to one another, which is handed down to each generation," write that "This sum of productive forces, capital funds and social forms of intercourse, which every individual and generation finds in existence as something given, is the real basis of what the philosophers have conceived as 'substance' and 'essence of man'" (1965, 50–51).

They go on to suggest that whereas Feuerbach's "conception" of the sensuous world is one of ahistoric contemplation by an "abstract" man, for the "real historical man," the natural world—even the objects of the simplest "sensuous certainty"—is never "given" but is mediated through culture and human praxis. Marx and Engels are clearly suggesting, in opposition to the contemplative materialists, that the essence of "man" is that she or he is a social being. Although some contemporary Marxists (e.g. Heather 1976) stress that these insights need to be incorporated into academic psychology—with its individualistic and behaviouristic bias—they are not specifically Marxist. The whole social scientific tradition has, over the last hundred or more years, been expressing this viewpoint in various ways, namely that what characterises humankind as a "species-being" is its social consciousness. This social consciousness mediates humankind's perception of nature and is an adaptive mechanism.[13]

What was specific to Marx and Engels was not only their stress on culture, but also their refusal to see "nature" and "history" as separate and antithetical. For them humans always had a "historical nature" (culture) and a "natural history" (a relationship with nature) (1965, 58). Hence, for Marx and Engels, what characterises human beings is their "practical-critical activity" (1968, 28–30). (The terms "practice" and "practical" are present in almost every one of the eleven *Theses on Feuerbach*.) As Bookchin put it, "Marx tried to root humanity's identity and self-discovery in its productive interaction with nature" (1982, 32). This implies that for Marx, "men" are both thinking and active beings, and that through labour they creatively produce their own "conditions of life." In turn, the mode of production of material life is seen to condition "the social, political, and intellectual life processes in general." And in that familiar phrase, Marx was to write: "It is not the consciousness of men that determine their being, but on the contrary, their social being that determines their consciousness" (1968, 181). Praxis, therefore is a key concept in Marxist theory. Marx counterposed Feuerbach's radical materialism and humanism with a notion of the human subject that conjoined theoretical and practical reason. For him, to speak of "mind" or "ego" as an abstract entity, unconnected with humanity was pure nonsense (see Lefebvre 1968, 3–58; Worsley 1982, 31) and this is why I think Piaget, and even Freud, are closer to the thoughts of Marx than Lévi-Strauss and most cultural anthropologists, and even some self-proclaimed Marxists.

In reacting against the "essentialism" of the early bourgeois theorists and following Althusser's polemics against Hegelian and humanistic Marxists, structural Marxists seem to suggest that the human "subject" is simply a creation of culture (or language). They thus join forces with both the cultural relativists (who are close to the German idealist tradition) and the Durkheimian sociologists in playing down the natural world, and the creative labour of the human subject. Although entitled *Language and Materialism*, Coward and Ellis's study of semiology and the "theory of the subject" is significant in that it hardly mentions the natural world, and makes no reference at all to the findings of Vygotsky, Whorf, Piaget, and Schaff. If, as Coward and Ellis seem to suggest, the "social process has no 'centre' or subject, or 'motivating force'" (1977, 74), and if the concept of "subject" is simply an "ideological construct" of the bourgeoisie, it is, I think, difficult to conceive how social life ever originated. And if it is conceded that "morality" depends on the "category of subject" for its functioning, then it is clearly problematic to postulate culture without a subject. Whereas the early bourgeois theorists—according to the premise of methodological individualism—had taken the human subject as given and "natural," and society as in some sense artificial and imposed, we now have the inversion of this: the "social system" is given (almost a "transcendental" entity) and the "subject" results from its construction in sociality. But the human subject and culture are dialectically related and coexistent. One can, of course, conceptualise the human subject abstracted from any social context—this is precisely the kind of theory that Coward and Ellis and Heather rightly, and cogently, criticise. Equally, one can treat culture (and language) as an abstract system without a subject, which is what Durkheim does. The danger is that one either collapses into cultural relativism or postulates an ahistoric human "mind," as does Lévi-Strauss and as is required by the structuralism of Althusser. It is not surprising, therefore, that Althusser, Durkheim and Lévi-Strauss have all been accused of following a Neo-Kantian, dualistic paradigm (Rose 1981, 1–39). But without a human subject (as defined by Marx) history is inevitably divorced from nature. Social structures are only prior to the subject if the latter is conceptualised as an organism, or as the idiosyncratic individual at birth, a viewpoint that Coward and Ellis try to counter. It would, of course, be highly misleading to equate the "abstract individual" of the early social contract theorists with the human "subject" of sociological analysis—as it would be to conflate empiricism with empirical data.

In seeing humankind as essentially social, and in putting a focal emphasis on productive relationships, Marx inevitably did not speak of individuals except in "so far as they are personifications of economic categories, representatives of special class relations and class interests" (1867, preface). And his key analytic category is not "culture" or "society" but "mode of production." To an important degree Marxist psychology implies a "class psychology" and a "collective" subject (see Bock 1980, 186–88; Anderson 1983, 44–45). But in seeing the world and emergent social life, in a Hegelian sense, as a kind of process, and culture as the creation of a human subject, Marx essentially attempted to transcend the familiar antinomies of nineteenth-century thought—idealism and materialism, theory and practice, hermeneutic understanding and analysis, history and nature, humanism and science and, in the present context, cultural relativism and universalism.

My own thoughts on the "theory of the subject" follow those of Lucien Goldman. Like Marx, Goldman rejects as untenable the many radical dualisms that pervade contemporary thought—philosophy and science, theory and praxis, interpretation and explanation. And on the concept of the subject he argues persuasively against the two main approaches. One gives the subject, specifically the *cogito*, analytical priority—a line of thought that stems from Descartes and is expressed by existentialists, phenomenologists, as well as by humanistic psychologists and interpretative sociologists. The other, characteristic of contemporary structuralism (and he specifically cites Althusser and Lévi-Strauss, whose theories Coward and Ellis closely follow), leads to the "negation of the subject."[14] Although the "subject" has the status of other scientific concepts in that it is constructed, Goldman holds that such a concept is a grounded one in having a necessary function of "rendering the facts we propose to study intelligible and comprehensible" (1977, 92). The first approach, which begins with the individual subject and puts a focal emphasis on meaning, is, Goldman argues, essentially non-explanatory and unable to account for the relationship among phenomena. The second approach, by negating the subject, is unable to account for the becoming or genesis of a structure, or its functioning. "The first does not see structure; the second does not see the subject which creates genesis, becoming and functionality" (1977, 106). Goldman, therefore, argues that it is necessary to integrate consciousness with behaviour and praxis, and to seek both the meaning and functionality of structures. To do this one needs a dialectic approach that situates a "creative subject at the interior of social life."

> To comprehend a phenomenon is to describe its structure and to isolate its meaning. To explicate a phenomenon is to explain its genesis on the basis of a developing functionality which begins with a subject. And there is no radical difference between comprehension and explication (1977, 106).[15]

But to argue against the theoretical negation of the human subject is not in the least to deny that the "self" or "mind" is socially constituted: the subject, human existence is no more conceivable outside social relations, than are social relations conceivable without subjects.

These issues have been discussed, in a number of interesting and perceptive essays, by A. Irving Hallowell, a much-neglected scholar, whose work has come to be recognised only in recent years. Through his students—Spiro, Bourguignon, and Wallace in particular—Hallowell has, however, had a deep and pervasive influence on anthropology. Although some earlier anthropologists such as Mauss, Radin, and Lee had discussed, from an anthropological perspective, the concepts of self and person, these concepts were generally neglected in ethnographic studies. Hallowell laid the foundations for the study of what has come to be termed "indigenous psychologies" (see Heelas and Lock 1981).

Hallowell observed that the evolution of the human species had been examined in terms of the taxonomy and phylogeny of the primates, the development of language or toolmaking, and social structure or culture as a specific human mode of adaptation. He suggested that an increased understanding of this complex process of human evolution might be gained in two ways. First we should get away from the idea that there is a radical discontinuity inherent in the evolutionary process, as if culture, language and "man" suddenly leapt into existence. He therefore postulated the concept of *protoculture*—a stage exemplified by non-hominid primates which entailed "simple forms of learning, some socialization of the individual, a social structure based on role differentiation in organized social groups, the transmission of some group habits and perhaps tool using and a 'non-syntactic' form of communication" (1960, 359; see also Hallowell 1976, 291–94; Bourguignon 1979, 29–39).

Second, he suggested a *conjunctive* approach to human evolution, seeing behaviour as "the unifying center" of the other significant variables. Toolmaking, for example, would then be interpreted as an early indication of the reality principle, involving ego functions. A "psychological dimension" could then be added to our conception of the personality structure of the early hominids. Thus Hallowell came to suggest that for hominid evolution to have advanced beyond the protocultural level, a major "psychological transformation" must also have occurred. The ecological development of our human forebears through the invention and use of technological devices, the normative orientation of human societies, involving regulations and moral precepts (such as incest), the cultural transmission of a human system of communication, all these Hallowell suggests, necessitated the existence of a concept of self, persistent in time. The ego permits adaptation at a new behavioural level. A capacity for self-awareness and self-identification must consequently be assumed as psychological potentialities. These would appear to be as necessary for the functioning of notions of eschatology as for the manufacture of tools and other forms of cultural adaptation (1976, 257). Thus Hallowell postulates that "without the development of self-awareness as an intrinsic part of the socialization process, without a concept of self that permits attitudes directed towards the self as an object to emerge and crystallize, we would not have some of the essential conditions necessary for the functioning of a human society" (1955, 83). Various points are worth noting about this perspective.

In placing a focal emphasis on the self and not on "mind," Hallowell, like Marx, conceptualises the human person as an essentially psychosomatic entity, as both an organic and a cultural being. Second, the self is seen as a constant factor in the human personality structure, and *intrinsic* to the operation of human societies, and all situations of social interaction. Self and society, for Hallowell, are aspects of a single whole, and culture and personality cannot be postulated as completely independent variables. Third, Hallowell suggests that neither a human society nor a human personality can be conceived in functional terms apart from systems of symbolic communication. Thus "social" existence was a necessary condition of the development of the self (or mind) in the individual. Likewise Hallowell argues that the development of the human psychological structure (mind, personality, self) is "fundamentally dependent upon socially mediated experience in interaction with other persons" (1955, 355). Like Goldman, Hallowell sees self and society as coexistent, and dialectically interdependent. Finally, although stressing the generic aspect of psychological structures, Hallowell also stressed that the nature of the self was itself a "culturally certifiable *variable*" (1955, 76). Human existence depended on making some distinction between "objective" and "subjective," for no person could nourish themselves if they lived in a dream world. But this categorical distinction, which is "one of the fundamental axes along which the psychological field of the human individual is structured for action in every culture" (1955, 84), is by no means unproblematic, for the simple reason that there is no one-to-one correspondence between experience and reality. "The psychological field in which human behaviour takes place is always culturally constituted, in part, and human responses are never reducible in their entirety to stimuli derived from an "objective" or surrounding world of objects in the physical or geographical sense" (1955, 84). In this Hallowell was echoing the tenet of a modified Whorfian theory and Marx's criticisms of Feuerbach's contemplative materialism. This meant that from the standpoint of the self the "environmental field" was culturally variable, and since the self is also partly a cultural product, the field of behaviour that is appropriate for the activities of *particular* selves in *their* world of culturally defined objects is not by any means precisely coordinate with any absolute polarity of subjective-objectivity that is definable (1955, 84).

With this notion of a generic personality structure—a structure which is fundamentally dependent on culturally mediated experience but which cannot simply be reduced to inherited organic structure—Hallowell's suggestions return us to those neglected founders of anthropology—Wundt, Boas, and Rivers—who all affirmed that biology and social science did not exhaust the theoretical options open to the human sciences. Indeed Marcel Mauss defined anthropology as the "sum total" of those sciences dealing with humankind (biology, psychology, sociology) which consider the person "as a living, conscious and sociable being" (1926, 5). The work of Hallowell readjusted the focus of anthropology to psychology, the central subject area of the human sciences.

Notes

Note: This essay was originally presented as a seminar paper at the Department of Anthropology, University of Edinburgh, in November 1984. I should like to thank those present for their helpful comments and criticisms. I am also grateful to Tim Ingold and Olivia Harris for criticising the first draft.

1   These two questions, although different, are closely interrelated. For there can be no true "psychology" without the notion of either the "psyche" or a conscious human subject to some degree independent of both the cultural and somatic domains, and to some extent influencing both.

2   For useful discussions of this issue, which was once an important debate, see Steven Lukes, "Methodological Individualism Reconsidered"; and Ernest Gellner, *Cause and Meaning in the Social Sciences*, 1–17. There is of course a fundamental distinction between the *abstract individual*, the *subject* of universal attributes postulated by early contract theorists and writers such as Leach, and the subject of existentialism and ethnomethodology, which is seen as creatively determining its own being. The first implies a form of crude naturalism while the voluntaristic emphasis of the second, if taken to extreme, denies the validity of any scientific approach per se to social theory. But the stress on voluntarism and subjectivity often leads, unconsciously, to bourgeois individualism. A good example is to be found in Mary Douglas, *In the Active Voice*. The "individual" (when he or she appears in an environment of "low grid") seems to be the very incarnation of the Hobbesian subject competitive, controlling, manipulating, autonomous, negotiating. Douglas seems to sit, as Louis Dumont ("Preface to the French edition of 'The Nuer'") might put it, in the "metaphysical armchair" of her own culture, universalising attributes that are culturally specific.

3   I am aware that there is a fundamental distinction to be made between sociobiology and behaviourism, and Durkheimian sociology. The kind of naturalistic reductionism suggested by sociobiologists would have been thoroughly anathema to Durkheim. Nevertheless both, in a sense, bypass the human subject.

4   Andrew Lock, following Hallowell's (1955) seminal thoughts on this issue, put it nicely when he suggested that "the concepts of self and culture are interdependent; one cannot exist without the other" (1981, 19). Ironically, whereas Althusser advocates a kind of Marxism which deals with ahistoric structures and, it would appear, leaves living people out of the analysis, other Marxists have criticised this kind of approach (exemplified by behaviourist psychology) as presenting the kind of "image of man" that is consonant with the needs of contemporary capitalism—the dehumanised person who is simply a reactive being (Ingelby, 1972). In a radical defence of a humanistic and historical conception of social science, Lucien Goldmann likewise pointed to the affinities between this kind of structuralism and the needs of advanced capitalism (1969, 13).

5   I hold that some sense can be made of social understanding by attempting to identify broad currents of thought—such as structuralism, cultural idealism, and behaviourism—but these categories are quite distinct from the perspectives of individual scholars whose writings may reflect several approaches. The approaches and phases I denote here, therefore, are tendencies and it would be ridiculous to use them to pigeonhole specific scholars.

6   One may debate how far Kroeber and Benedict should be described as "functionalists," but the important point is that Kardiner saw an affinity between Benedict's "integrated whole," Kroeber's "super-organic," and Malinowski's functionalism—and was critical of all three to the degree that they excluded the world of "flesh," "bone," and the "individual."

7   Biographers of Reich such as Wilson and Rycroft have shown little sympathy for either the man or his ideas; for an alternative view see Sharaf (1983).

8   The concept of the "unconscious" as such does not play a major role in Reich's theorising, especially when he is compared with Freud and Jung. Jung's concept of a "collective unconscious" is positive, but it has mystical overtones; Freud's unconscious,

on the other hand, though grounded in biology, is seen in negative terms. Reich, in contrast to both, thought of the "unconscious" as a sexual force which, if allowed to discharge freely through genital love, was wholly positive in its effects. The limitations of Reich's biologism and his hydraulic theory of sexuality have been stressed by many feminist writers.

9   In these paragraphs I am concerned only with indicating Reich's research strategy, not with assessing the validity of his analysis of fascism, see Kitchen (1976, 13–14), Poster (1978, 46–52), Laclau (1979, 84–85).

10   With Fromm, evolution and history are in a sense equated. His approach to history is teleological and has a mythical quality. The human subject, having historically emerged with the breakdown of feudalism, is faced with a moral alternative: either continue to embrace an authoritarian culture or reunite with nature and other humans in love and mutual productivity. It is probably at this kind of humanism that Althusser's critique is aimed.

11   Schaff (1963) explores the inherent contradiction in existentialism between its "philosophy of despair" and its stress on the human individual, creatively determining his or her own being.

12   Durkheim of course, unlike Hobbes and later methodological individualists, gave "society" priority, arguing that it was not so much a human artefact as a reality *sui generis*. See Lukes (1973) and Rose (1981, 14–18), both of whom stress Durkheim's dualism.

13   These are complex issues which I cannot pretend to have unravelled, but, as Tim Ingold has suggested to me, the whole social scientific tradition has been divided over at least the last century on the question of whether consciousness reflects an internalised cultural logic, or whether culture is the vehicle or instrument of consciously directed human practice. Thus a distinction may be made between postulating that the "essence" of humanity is objective culture—the "symbolic"—internalised in each individual prior to the interactive process, and that the "essence" of humanity is consciousness as a creative movement that enfolds, and unfolds in, social relations. My suggestion is that Durkheim, writers in the German idealist tradition such as Boas and Benedict, and structural Marxists follow the first tendency, while Marx tends to adopt the second—with the proviso that he follows Hegel in seeing sociality as expressed only in relation to an "objective" natural world.

14   I do not wish to dismiss the important work of Althusser and Lévi-Strauss. Althusser's critique of socialist humanism was a necessary corrective to the voluntarism implied by the early Sartre and the romantic-irrationalist attitude towards science that was often expressed by phenomenological Marxists (see Hirsh 1981 and Benton 1984). Althusser simply went to the other extreme and implied that human agents were simply bearers of social structures and contrary to what Marx had suggested, did *not* make their own history. Likewise, although Lévi-Strauss was a severe critic of phenomenology and existentialism with its "illusions of subjectivity" (as his own polemics with Sartre denote), it is important to note that these shafts of criticism were not specifically aimed at human agency per se, but rather at the Cartesian notion that human consciousness has priority in cultural understanding. Like Hegel, Marx, and Freud he is anti-empiricist rather than anti-humanist. Indeed, in stressing the priority of the "lived-in" orders, Lévi-Strauss appears, rather like Kardiner, to place the operations of the unconscious mind as a kind of mediator between praxis (the economic infrastructures) and cultural beliefs and practices (the "thought-of" orders). But whereas Kardiner's basic personality structure is culturally specific, Lévi-Strauss's "mind" is pan-human.

15   Recent articles by Milton Singer (1978, 1980), dealing with linguistic theory, offer a very similar perspective to that of Goldman. In arguing against Saussure's semiology which defines the sign function as a dyadic relation of the signifier (sound-image) and signified (concept)—thus dispensing with independent objects and subjects— Singer suggests that Peirce's semiotic theory offers a much better foundation for psychology and cultural anthropology. For in combining an anti-Cartesian and phe- nomenological approach with pragmatism, and in basing his theory on an irreducible triad of object, sign, and interpretant, Peirce was able, Singer argues, to produce a semiotic theory of the self that was consistent with both science and common sense. Avoiding both subjectivism and objectivism (Saussure's semiology), Peirce is seen to postulate the self as both the object and subject of semiotic systems. But whether this kind of anti-Cartesian dualism is "Kantianism with an empirical subject" (as Singer suggests) is questionable, for in Kant's view the objective world—the "thing in itself"—can be sensed but never known.

References

Allport, Gordon W. 1955. *Becoming*. New Haven: Yale University Press.

Althusser, Louis. 1972. *Politics and History*. London: NLB.

Anderson, Perry. 1983. *In the Tracks of Historical Materialism*. London: Verso.

Bateson, Gregory, and Margaret Mead. 1942. *Balinese Character*. New York: Academy of Sciences.

Benton, Ted. 1984. *The Rise and Fall of Structural Marxism*. London: Macmillan.

Bock, Philip K. 1980. *Continuities in Psychological Anthropology*. San Francisco: W.H. Freeman.

Bookchin, Murray. 1982. *The Ecology of Freedom*. Palo Alto, California: Cheshire.

Bourguignon, Erika. 1979. *Psychological Anthropology*. New York: Holt, Rinehart.

Brown, Phil. 1974. *Toward a Marxist Psychology*. New York: Harper & Row.

Callinicos, Alex. 1976. *Althusser's Marxism*. London: Pluto Press.

Coward, Rosalind, and John Ellis. 1977. *Language and Materialism*. London: Routledge & Kegan Paul.

Devereux, George. 1978. "The Works of George Devereux." In *The Making of Psychological Anthropology*, edited by George D. Spindler. Berkeley: University of California Press.

Douglas, Mary. 1982. *In the Active Voice*. London: Routledge & Kegan Paul.

Dumont, Louis. 1975. "Preface to the French Edition of 'The Nuer.'" In *Studies in Social Anthropology*, edited by J.H.M. Beattie and R.G. Lienhardt. Oxford: Clarendon Press.

Durkheim, Émile. 1915. *The Elementary Forms of the Religious Life*. London: Allen & Unwin.

_____. 1974. *Sociology and Philosophy*. New York: Free Press.

Fromm, Erich. 1942. *The Fear of Freedom*. London: Routledge & Kegan Paul.

_____. 1949. *Man for Himself*. London: RICP.

_____. 1962. *Beyond the Chains of Illusion*. New York: Sphere Books.

_____. 1970. *The Crisis of Psychoanalysis*. Harmondsworth: Penguin.

Gellner, Ernest. 1973. *Cause and Meaning in the Social Sciences*. London: Routledge & Kegan Paul.

Goldmann, Lucien. 1969. *The Human Sciences and Philosophy*. London: J. Cape.

_____. 1977. *Cultural Creation in Modern Society*. Oxford: Blackwell.

Hallowell, A. Irving. 1955. *Culture and Experience*. Philadelphia: University of Pennsylvania Press.

_____. 1960. "Self, Society and Culture in Phylogenetic Perspective." In *Evolution after Darwin*, edited by Sol Tax. Chicago: University of Chicago Press.

_____. 1976. *Contributions to Anthropology*. Chicago: University of Chicago Press.

Harris, Marvin. 1969. *The Rise of Anthropological Theory*. London: Routledge & Kegan Paul.

Heather, Nick. 1976. *Radical Perspectives in Psychology*. London: Methuen.

Heelas, Paul, and Andrew Lock, eds. 1981. *Indigenous Psychologies*. New York: Academic Press.

Hirsh, Arthur. 1981 *The French New Left*. Boston: South End Press.

Hirst, Paul, and Penny Woolley. 1982. *Social Relations and Human Attributes*. London: Tavistock.

Ingelby, David. 1972. "Ideology and the Human Sciences." In *Counter Course*, edited by Trevor Pateman. Harmondsworth: Penguin.

Jay, Martin. 1973. *The Dialectical Imagination*. London: Heinemann.

Kaplan, David, and Robert A. Manners. 1972. *Culture Theory*. New York: Prentice Hall.

Kardiner, Abram. 1939. *The Individual and His Society*. New York: Columbia University Press.

_____. 1945. *The Psychological Frontiers of Society*. New York: Columbia University Press.

Kardiner, Abram, and Edward Preble. 1961. *They Studied Man*. New York: Mentor.

Kitchen, Martin. 1976. *Fascism*. London: MacMillan.

Kluckhohn, Clyde. 1944. *Navaho Witchcraft*. Boston: Beacon Press.

Köhler, Wolfgang. 1937. "Psychological Remarks on Some Questions of Anthropology." *American Journal of Psychology* 50: 271–88.

Laclau, Ernesto. 1979. *Politics and Ideology in Marxist Theory*, London: Verso.

Laing, R.D. 1967. *The Politics of Experience*. Harmondsworth: Penguin.

Lefebvre, Henri. 1968. *The Sociology of Marx*. Harmondsworth: Penguin.

LeVine, Robert A. 1973. *Culture, Behaviour and Personality*. Chicago: Aldine.

Lewis, I.M., ed. 1977. *Symbols and Sentiments*. New York: Academic Press.

Lock, Andrew. 1981. "Universals in Human Cognition." In *Indigenous Psychologies*, edited by P. Heelas and A. Lock. New York: Academic Press.

Lukes, Steven. 1970. "Methodological Individualism Reconsidered." In *Sociological Theory and Philosophical Analysis*, edited by Dorothy Emmet and Alasdair MacIntyre. London: Macmillan.

_____ 1973. *Émile Durkheim: His Life and Work*. Harmondsworth; Penguin.

Marx, Karl. 1959. *Economic and Philosophic Manuscripts of 1844*. Moscow: Progress.

_____. 1974 (1857). *Capital*, Vol. 1. London: Dent.

Marx, Karl, and Friedrich Engels. 1965. *The German Ideology*. London: Lawrence & Wishart.

_____. 1968. *Selected Works*. London: Lawrence & Wishart.

Mauss, Marcel. 1926. *Sociology and Psychology*. London: NLB.

Mitchell, Juliet. 1975. *Psychoanalysis and Feminism*. Harmondsworth: Penguin.

Novack, George. 1966. *Existentialism Versus Marxism*. New York: Dell.

_____. 1973. *Humanism and Socialism*. New York: Pathfinder.

_____. 1978. *Polemics in Marxist Philosophy*. New York: Monad Press.

Piaget, Jean. 1971. *Structuralism*. London: Routledge & Kegan Paul.

Poster, Mark. 1978. *Critical Theory of the Family*. London: Pluto Press.

Reich, Wilhelm. 1929. "Dialectical Materialism and Psychoanalysis." London: Socialist Repr.

_____. 1930. *The Sexual Revolution*. London: Vision Press.

_____. 1932. *The Invasion of Compulsory Sex-Morality*. Harmondsworth: Penguin.

_____. 1933. *The Mass Psychology of Fascism*, Harmondsworth: Penguin.

_____. 1942. *The Function of the Orgasm*. London: Panther.

Robinson, Paul A. 1970. *The Sexual Radicals*. London: Paladin.

Rose, Gillian. 1981. *Hegel Contra Sociology*. London: Athlone.

Schaff, Adam. 1963. *A Philosophy of Man*. New York: MRP.

Shalvey, Thomas. 1979. *Claude Lévi-Strauss*. Brighton: Harvester Press.

Sharaf, Myron R. 1983. *Fury on Earth*. London: Hutchinson.

Singer, Milton. 1978. "For a Semiotic Anthropology." In *Sight, Sound, and Sense,* edited by
　　Thomas A. Sebeok. Bloomington: Indiana University Press.
———. 1980. "Signs of the Self: An Exploration of Semiotic Anthropology." *American
　　Anthropologist* 82: 485–507.
Stirner, Max. 1845. *The Ego and His Own.* New York: Dover.
Thompson, E.P. 1978. *The Poverty of Theory.* London: Merlin.
Voget, Fred W. 1975. *A History of Ethnology.* New York: Holt, Rinehart.
White, Leslie A. 1949. *The Science of Culture.* New York: Grove.
Wilson, Colin. 1981. *The Quest for Wilhelm Reich.* London: Granada.
Worsley, Peter. 1982. *Marx and Marxism.* London: Tavistock.

2

In Defence of Realism and Truth: Critical Reflections on the Anthropological Followers of Heidegger (1997)

Prologue
Postmodernism is now all the rage in anthropology. Scholars who only a decade or so ago were making a fetish out of science or Marxism now seem to be repudiating them entirely, and have embraced postmodernism with an uncritical fervour. This article offers some critical reflections on this current Heideggerian vogue.

Nobody seems to know exactly what *is* postmodernism, and one anthropologist even admits guilt feelings about using the term (Fabian 1994, 103)—but its effect on anthropology, whose essential insights it has in fact largely expropriated, has been quite remarkable. Anthropology has never been a monolithic discipline. It has always reflected a diverse range of theoretical perspectives, but seen as a critical movement within anthropology (past anthropology being depicted—or caricatured—as "positivist") postmodernism seems to be creating a "sense of disarray" within the discipline. We are now in a state of "crisis," of epistemological "turmoil," and postmodernism, we are informed, has completely shattered the foundations of anthropology, seriously questioning the legitimacy of studying other cultures (Nencel and Pels 1991, Hastrup 1995).

In addition, the anthropology of the past is now portrayed as being simply the study of the "primitive," the "exotic" other, and was thus engaged largely in a kind of "salvage" operation of "disappearing" cultures. This seems a rather biased and misleading portrait of anthropology (see Vincent 1990, Goody 1995), for the discipline has a long tradition of "anthropology at home" and it is none too clear how the ethnographies of Embree, Aiyappan, Kenyatta, Srinivas, and Fei fit into this portrayal of anthropology. It is of course noteworthy that James Clifford and George Marcus (1986), in what many have regarded as the founding text of postmodernist or literary anthropology, are not only rather dismissive of feminist anthropology but ignore completely the ethnographic studies of non-Western scholars.

At the present juncture, however, some anthropologists have embraced postmodernism and its rather nihilistic vision with an uncritical and unbounded enthusiasm; while others have denounced it in rather harsh terms as if it were a kind of disease. Given the reaffirmation of a more scientific—positivist—anthropology, drawing its inspiration from evolutionary biology, we now have a regrettable state of affairs where postmodernist (literary, hermeneutic) and positivist (sociobiology, empiricist) anthropologists are hurling disparaging epithets at each other. The hopes of Enlightenment thinkers like Vico, and other classical scholars (Marx, Dilthey, Freud, Weber) of combining humanism (hermeneutics) and naturalism (science) as a poetic or humanistic science, seems to be further away from achievement than ever before. For post modernists, empirical science is *equated* with capitalism, with state power, and with the technological mastery of nature—and thus repudiated entirely—while sociobiologists and those who seem themselves as "defenders" of the Enlightenment, pour scorn on hermeneutics as if it was some malady, or a self-indulgent "delusion" (Tyler 1986, Gellner 1995). The whole idea of being for or against the Enlightenment as a period, or as a cultural ethos (in all its diversity), is completely ahistoric and unhelpful. No wonder Foucault, who had a sense of both history and truth, refused to succumb to this kind of "blackmail"— as he called it (Rabinow 1984, 42).

There is much that is valid and important in the postmodernist critique—as there is in the writings of Nietzsche and Heidegger, from whom the postmodernist philosophers (Foucault, Derrida, Lyotard, Rorty) derive—selectively—most of their essential premises. Postmodernist anthropologists, suffering from a kind of amnesia with regard to the roots of their own discipline—which has always emphasised "difference" (the diversity of local cultures) sometimes to the point of absurd exaggeration, and which is inherently reflexive (Hastrup 1995, 49–50)—seem to follow somewhat mesmerised in the wake of these philosophers. The emphasis on the historicity of being; the critique of Cartesian metaphysics and the transcendental subject; the undermining of the dualistic opposition between humans (culture) and nature; the importance of hermeneutic understanding; the stress that there is no unmediated relationship between language (or consciousness) and the world; the notion that social experience (the life world) forms the basis and background of any theoretical standpoint (that humans are both practical and contemplative beings); the problematic nature of instrumental reason and the problems of equating truth with science—all these are important issues highlighted by postmodernist anthropologists. However, anxious to affirm their own intellectual importance and originality, many postmodernist and literary anthropologists seem to suffer from historical amnesia and forget that all of these issues have been articulated for more than a century, by numerous people—Neo-Kantian scholars like Dilthey, Wundt, and Boas; evolutionary biologists; Marx and those of his followers who have stayed close to the Hegelian tradition; pragmatic philosophers like Dewey and Mead; and social scientists more generally. Indeed, as Derek Layder suggests (1994, 17), the inherently social nature of human life has been taken for granted as a fundamental premise of the social sciences for the past hundred years, including anthropology. Anthropology, indeed, has always included an implicit critique of what has variously been described as "metaphysics" (Heidegger), "logocentrism" (Derrida), and "epistemology" (Rorty). When Stephen Toulmin (1995) recently contended that a "radical challenge" to Cartesian metaphysics—involving an emphasis on the social nature of knowledge, on the fact that there is no certain knowledge, and that the mind is not a passive container, a camera obscura—only came to be articulated during the last thirty years one can only gasp at such myopia. But it is the common scenario of many postmodernists who assume that nothing of intellectual importance happened in the three hundred years between 1650 and 1950! In my study on Western conceptions of the individual (1991) I reviewed the long history of the many challenges to the "Cartesian programme"—which is usually equated with the Enlightenment, although, in fact, it was the Enlightenment thinkers who first initiated the critique of this programme and its ahistoric rationalism.

This exaggerated reaction against Cartesian metaphysics and positivism (which tend to be conjoined) by Tyler (1986, 1991), Flax (1990), and other postmodernists, invokes a rather bleak scenario, where the choice we are given is between two extremes: absolutism or nihilism. Thus we have the following theoretical moves:

1. As we have no knowledge of the world except through "descriptions" (to use Rorty's term) the "real" is conceived only as an "effect" of discourses. As Derrida rhetorically put it: "there is nothing outside of the text" (1976, 158). *Realism* is thus repudiated.

2. As there is no unmediated relationship between consciousness (or language) and the world—to assume a one-to-one correspondence is dubbed "logocentrism," the "metaphysics of presence," or the "mirror" conception of knowledge—postmodernists go to the other extreme and posit no relationship at all between language and the world—or if there is, it is completely indeterminate. Truth is thus either repudiated entirely (as with Tyler), or seen as simply an "effect" of local

cultural discourses (as with Flax), or it is seen as something that will be "disclosed" or "revealed" by elite scholars through poetic evocation (as with Heidegger). *Truth* as correspondence or *representation* is thus repudiated.

3. As the transcendental ego of Cartesian rationalism (or Husserl's phenomenology)—the epistemic subject standing outside of both history and nature—is disavowed by postmodernists (it has of course been critiqued for more than a century), they go, like the structuralists, to the other extreme, and announce the complete dissolution of human subjectivity and agency. The *self* is thus repudiated.

4. As there are no absolute truths, for neither the transcendental ego nor empirical sense-data provide us with firm "foundations" for epistemological "certitude," postmodernists like Tyler and Flax go to the other extreme, and affirm that all "knowledge" (the term in fact becomes redundant!) is subjective, fragmentary, undecidable, indeterminate, relative, or contingent. Both objective *knowledge* and *empirical science* are thus repudiated.

Postmodernist scholars exclaim with some stridency, the "dissolution," the "erasure," or the "end" of truth, reason, history, nature, the self, science, and philosophy—misleadingly identifying all these terms with conceptions that are transcendental, ahistoric, and absolutist. They thus appear to see nothing between the so-called "god's eye" point of view, a transcendental perspective beyond time and space, and local—supposedly fragmented, undecidable, and indeterminable—discourses (Hollinger 1994, 81). In the process, a sense of common humanity, of human capacities, of human praxis, of human history is lost. There is no sense of a human life-world—"an infinitive surrounding world of life" common to all people, as Husserl expressed it (1970, 139), that is prior and distinct both from cultural worldviews and transcendentalism. Yet the postmodernists nevertheless recoil from the theoretical implications of their own rather prophetic declarations, and with equal emphasis proclaim that their "theory" does not entail either linguistic (cultural) idealism, or relativism (Hollinger 1994, 98; Flax 1995, 155). They could hardly do otherwise, for outside the groves of academia, and their reified and "scholastic" discourses, the natural and social worlds are experienced as a reality, and we experience also a shared humanity that is not reducible to the fragmented discourses of local cultures.

In this article, contrary to the nihilistic ethos of postmodernism, I wish to affirm the salience of a realist metaphysics, and the crucial importance of such conceptions as truth and representation, human agency and empirical knowledge. I will discuss each of these four issues in turn.

Realism

It has long been known, of course, long before postmodernism came upon the anthropological scene, that we do not perceive or experience the world in pristine fashion. For our engagement with the world is always mediated by our personal interests, by our state of mind, by language and cultural conceptions, and, above all, by social praxis. As an early and important historian of science put it in 1838; "there is a mask of theory over the whole face of nature" (Megill 1994, 66). The anthropologist Ruth Benedict in her classic anthropological text *Patterns of Culture* long ago emphasised that a person's ideas, beliefs, and attitudes are largely culturally constituted. As she wrote: "No man ever looks at the world with pristine eyes. He sees it edited by a definitive set of customs and institutions and ways of thinking" (1934, 2).

But as with Dilthey, her important mentor, this affirmation did not in the least imply a denial of the reality of the material world. This important insight, which has been part of the common currency of the social sciences ever since the time of Marx, has, in recent decades, been taken up by philosophers and postmodernist anthropologists. But they seem to have taken this important insight to extreme, and in a "veritable epidemic" of "social constructivism" and "world making" have propounded a latter-day version of Kantian idealism, going even further than Kant in denying the reality of the material world, the "things-in-themselves." Cultural idealism in its various guises is thus now all the rage in the halls of academia, and has been adapted by a wide range of scholars—Kuhn, Althusser, Goodman, Rorty, Hindess and Hirst, and Douglas, as well as postmodernist anthropologists (Devitt 1984, 235). Such constructivism combines two basic Kantian ideas: that the world as we know it is constituted by our concepts, and that an independent world is forever beyond our ken (ix). But, as said, many anthropologists go even further in an anti-realist direction and deny the independent existence of a world beyond our cognition, a world that has causal powers and efficacy. With the free use of the term "worlds" they invariably conflate the cognitive reality which is culture—"discourses" is now the more popular term—and the material world that is independent of humans. Thus anthropologists now tell us that there is "no nature, no culture," or that nature, sex, emotions, the body, the senses, are purely social "constructs" or human "artefacts," or even that they do not "exist" outside of Western discourses. The suggestion that "nature" is a human construct or artefact (rather than being simply constituted or "edited"), or that it has "disappeared" or does not "exist" are highly problematic notions. Derrida writes that "nature, that which words . . . name, have always already escaped, have never existed" (1976, 159). Although making an important point about the nature of language, this phrase simply indicates just how alienated from nature contemporary philosophers seem to be. They seem unable to recognise that human beings are, as Nigel Pennick puts it "rooted in the earth" (1996, 7). In a gleeful phrase, John Passmore notes that it is the French intellectual's dream "of a world which exists only in so far as it enters into a book" (1985, 32).

Now one means by "nature" the existential world in which we find ourselves—the trees, the clouds, the sky, the animals and plants, the rocks, and all those natural processes which are independent of human cognition, and on which human life depends. To suggest that this a human creation or artefact, or does not exist, is plainly absurd. Or, on the other hand, one means by nature the highly variable "concept" of nature; to suggest that this is a social construct is rather banal, though the suggestion is dressed up as if it was some profound anthropological insight. Reacting against the notion that there is an isomorphic—reflective—relationship between consciousness (language) and the world—so-called "logocentrism"—post-modernist anthropologists now seem to embrace a form of cultural (or linguistic) idealism, and deny the reality of the material world (nature), or sex or the senses. Although some anthropologists deny that "sex" exists (as there is *nothing* "pre-social," Moore 1994, 816–19 affirms) baboons in Malawi have no difficulty at all in distinguishing between male and female humans! (see Caplan 1987 for a more balanced perspective.)

Realism, as many philosophers have insisted, is a metaphysical doctrine. It is about what exists in the world and how the world is constituted. Contrary to what Kirsten Hastrup writes (1995, 60), it is not a theory of knowledge, or of truth, but of being, and so does not aim at providing a "faithful reflection of the world." Realism, as a doctrine, is thus separate from semantic issues relating to truth and reference, and from issues dealing with our knowledge of the world (epistemology)—both human and natural. Roy Bhaskar has critiqued what he describes as the "epistemic fallacy," the notion that ontological issues can be reduced to, or analysed in terms of statements about knowledge (epistemology) (1989, 13).

Everyone, of course, is a "realist" or "foundationalist" in some sense, making ontological assumptions about what is "real" and what "exists." Metaphysics is thus not something that one can dispense with, or put an "end" to—as both positivists and Heidegger and his acolytes suggest. For Plato, "ideas" or universals were "real"; for Descartes, the transcendental ego was "real"; for some eco-feminists the mother goddess is "real"; while for empiricists it is sense impressions. As used here, realism entails the view that material things exist independently of human sense experience and cognition. It is thus opposed to idealism which either holds that the material world does not exist, or is simply an emanation of spirit, or that external realities do not exist apart from our knowledge or consciousness of them. Outside of philosophy departments, and among some religious mystics and anthropologists, realism is universally held by everybody, and forms the basis of both common sense and empirical science. Common sense, of course, *sensus communis*, can be interpreted in Aristotelian fashion as a kind of sixth sense that draws together the localised senses of sight, touch, taste, smell, and hearing. It is this sixth sense, as Arendt writes, that gives us a sense of realness regarding the world (1978, 49). Like Arendt, Karl Popper critically affirmed the importance of common sense. He wrote: "I think very highly of common sense. In fact, I think that all philosophy must start from commonsense views and from their critical examination." But what, for Popper, was important about the common sense view of the world was not the kind of epistemology associated with the empiricists—who thought that knowledge was built up out of sense impressions—but its realism. This is the view, he wrote "that there is a real world, with real people, animals and plants, cars and stars in it. I think that this view is true and immensely important, and I believe that no valid criticism of it has ever been proposed" (Miller 1983, 105).

Science therefore was not the repudiation of commonsense realism, but rather a creative attempt to go beyond the world of ordinary experience, seeking to explain, as he put it, "the everyday world by reference to hidden worlds." In this it is similar to both religion and art. What characterises science is that the product of the human imagination and intuition are controlled by rational criticism. "Criticism curbs the imagination but does not put it in chains" (1992a, 54). Science is therefore, for Popper, "hypothetico-deductive."

What exists, and how the world is constituted, depends, of course on what particular ontology or "world view" (to use Dilthey's term) is being expressed, although in terms of social praxis the reality of the material world is always taken for granted for human survival depends on acknowledging and engaging with this world. As Marx expressed it, we are always engaged in a "dialogue with the real world" (1975, 328).

It is important then to defend a realist perspective, one Marx long ago described as historical materialism. It is a metaphysics that entails the rejection both of contemplative materialism (the assumption that there is a direct unmediated relationship between consciousness [language] and the world) and constructivism. The latter is just old-fashioned idealism in modern guise, the emphasis being on culture, language, and discourses, rather than on individual perception (Berkeley) or a universal cognition (Kant). This approach may also be described as dialectical naturalism (Bookchin 1990), transcendental or critical realism (Bhaskar 1978, 25; Collier 1994), or constructive realism (Ben-Ze'ev 1995, 50)—recognising the significant social and cognitive activity of the human agent, but acknowledging the ontological independence and causal powers of the natural world. As Mark Johnson simply puts it: "How we carve up the world will depend both on what is "out there" independent of us, and equally on the referential scheme we bring to bear, given our purposes, interests, and goals" (1987, 202). Our engagement with the world is thus always mediated. Equally important is the fact that we are always, as Marx put it, engaged in a "dialogue" with the material world. It is thus necessary to reject both idealism (constructivism) and reductive materialism (positivism, objectivism) as many classical sociologists and human scientists have insisted (see my *Anthropological Studies of Religion*, 1987). Again, Johnson expresses this rather well: "Contrary to idealism, we do not impose arbitrary concepts and structure upon an undifferentiated, indefinitely malleable reality— we do not simply construct reality according to our subjective desires and whims. Contrary to objectivism, we are not merely mirrors of nature that determines our concepts in one and only one way" (207). The "end" of metaphysics is, of course, simply an intellectual posture of the positivists and the Heideggerians, for we all affirm in our beliefs and writings certain ontological assumptions about the world. What Flax means by the "end" of "metaphysics" is a rejection of a certain kind of idealist, or absolutist metaphysics, one of course, that the social sciences rejected long, long ago.

Truth and Representation

The classical theory of truth as correspondence or representation has been taken for granted by ordinary people of all cultures and by most philosophers from Aristotle to Davidson. Knowledge consists of the search for truth. It is a form of cognition shared by science and ordinary common sense, and, as Arendt suggests, is quite distinct from thinking, or philosophy, which goes beyond what is known, and involves a search for meaning (often interpreted as universal truths). The distinction between knowing (truth) and thinking (meaning) is akin to Aristotle's distinction between *theoria* (rational contemplation or scientific knowledge) and *sophia* (philosophic wisdom, which combines science with intuitive reason—a knowledge of first principles). It is akin too, to Kant's distinction between *verstand*, intellectual knowledge, and *vernunft*, dialectical reason. Nothing but confusion reigns if science and philosophy are equated, and, as with the positivists and Heidegger, a "basic fallacy" is committed by interpreting "meaning on the model of truth" (Aristotle 1925, Book 6; Arendt 1978, 14–19).

The classical definition of truth was expressed by medieval scholars as: *veritas est adaequatio rei et intellectus*, "truth is the agreement of knowledge with its object." That this is what constitutes truth—not, as with coherence and pragmatic theories, what is the criterion of truth—has been acknowledged by philosophers throughout the ages: Aristotle, Ibn Sina, Aquinas,

Kant, Husserl, Popper, Davidson (see Husserl 1970, 176; Davidson 1984, 37; Popper 1992, 5). Kant noted that the definition of truth was expressed by the words "the accordance of the cognition with its object" but then went on to suggest that looking for a universal or secure criterion of truth, was a bit like one person "milking the he-goat, and the other holding a sieve" (1934, 67). Needless to say, the very idea of making a mistake or an error presupposes a realist metaphysic and the classical conception of truth (Trigg 1980, xx).

It is misleading, indeed obfuscating, to conflate, as many postmodernists do, the correspondence theory of truth (as representation) with a "objectivist" theory of meaning which assumes an isomorphic relationship between consciousness (language) and the world. It is equally obfuscating to equate the correspondence theory of truth with "absolute truth" (see Carrithers 1992, 153). The idea of a transparent, unmediated relationship between linguistic expressions and the world, as implied by early classical philosophers, especially Locke, and by the positivists, e.g. the "picture theory" of the early Wittgenstein—has been critiqued as "logocentrism" (Derrida), or as implying the notion that the mind is a "mirror of nature" (Rorty 1979, 170), or more recently, as "objectivism" (Johnson 1987). But, as many scholars have insisted, the classical theory of truth as correspondence does not entail this crude notion of resemblance or mirroring, such that truth, or scientific knowledge is simply a reflection of the natural (or social) world (Danto 1968, 144; Devitt 1984, 50; Collier 1994, 239–41). Truth as representation does not imply an isomorphism between a linguistic expression (or theoretical account) and the world that is being described. And it is quite misleading to imply that because words get their meaning from their relations (of contrast) with other words within a system of meaning structures, there is then no relationship between language and an extra-linguistic reality, as Derrida and his followers seem to insist. If consciousness (language) and the world already coincide and are isomorphic it could hardly be asked whether or not they correspond (Collier 1985, 195); or as Marx put it on a related issue, "science would be superfluous if the outward appearance and the essence of things directly coincided." If the correspondence theory of truth implied isomorphism it is also difficult to understand why scientists conduct elaborate experiments or why people spend time gathering evidence to substantiate their claims to truth.

Both Arendt and Popper insist that knowledge involves a search for truth, and truth consists of "the correspondence of knowledge with its object." But since we cannot know anything for sure, "it is simply not worth searching for certainty; but it is well worth searching for truth" (Popper 1992, 4)—or as Arendt puts it "provisional verity" (1978, 59).

But in the classical theory of truth as correspondence, truth neither finds its "essential locus" in the statements (as Heidegger seems to suggest

[Krell 1978, 122]), nor is truth out there in the world, existing independently of the human mind (as Rorty [1989, 5] seems to imply); but rather it
consists of a relationship between descriptions and the world. As Arthur
Danto clearly expressed it: "Truth is not a property of the world. It is not
a property of sentences either. Truth belongs neither to language nor to
the world, but to the relationship between them. A sentence is true when
it corresponds to the world." Thus "truth" describes nothing in the world,
but, as he puts it pertains to the "space" between the world and language"
(1968, 144).

A rejection of logocentrism—interpreted as implying that language
is a mirror or reflection of things in the world by natural resemblance
(Derrida 1976, 11)—does not entail a complete rejection of reference, representation, or truth as correspondence (Assiter 1996, 60–66). That language
is not anchored to the world by a logocentric one-to-one correspondence,
does not imply that there is no relationship between language and the
world. Surely, as Iris Murdoch writes in her powerful critique of the "heroic
aestheticism" of Heidegger and Derrida, words have definite meanings
when we apply them in particular social contexts (1992, 200). But semantic
theories of meaning ought not to be conflated with the notion of truth as
correspondence or representation. One can thus reject logocentrism or
"objectivism"—the idea that there is an unmediated relationship between
language and the world—without this entailing a rejection either of a
realist metaphysic or the correspondence theory of truth (Johnson 1987,
200–212). If one wants to know whether it is indeed true that the cat is
sleeping on the bed, one must have both prior understanding to be able to
recognise the cat, and engage in a practical activity to ascertain whether
the proposition does in fact accord with a state of affairs in the world. As
philosophers of science (Quine) and anthropologists have long recognised,
there is no direct, unmediated relationship between language (scientific
theories) and the world. But to suggest that there is no "transcendental"
signified, no world to which language can have any reference, is pure linguistic idealism. This is how Murdoch rightly interprets Derrida's philosophy, and his theory of "archi-écriture" (primal writing), which she considers to consist largely of "dramatised half-truths and truisms" (1992, 185–88).
The same can be said of much of Heidegger's philosophy.

Knowledge as truth, and as the representation of some given object
of study—such as hunting in Malawi, or local cultural schemas—entails
of course, re-presentation, the making present of what is actually absent.
This involves a unique gift of the human mind—imagination (Arendt 1978,
76). The notion that scientific thought does not involve the imagination is
one of the popular misconceptions of science—but science also involves
the critical testing of evidence for a particular theory, which thus makes it

distinct from poetry (Medawar 1982). When Hastrup suggests that anthropology consists "not of representations but of propositions about reality" (1995, 45), though acknowledging a realist perspective, she misleadingly defines the first term—representation—as implying an isomorphic or "mirror" image of the world. This leads her to repudiate the "realist" monograph—which allegedly represents societies as "timeless, island-like entities"—and to perceive representation as a "creative process of evocation and re-enactment" (1995, 21). If by "evocation" Hastrup means the representation of some given social reality that can be critically scrutinised by others, particularly by members of the community involved, then this seems little different from what social scientists and ethnographers have been doing for a long time. If, however, by "evocation" she means the disclosure of truth in the manner of Heidegger (that is, simply to engage in hermeneutics) or as a form of poetic meditation that describes "nothing" and is "beyond truth" (in the fashion of Tyler) then such a strategy can hardly provide us with valid knowledge. Hastrup's enlivening study is bedevilled throughout by the conflation of "realism" and "correspondence theory" with positivism, and with an "objectivist" theory of meaning that misleadingly assumes that "representation" necessarily entails an isomorphic, unmediated relationship between a theoretical account and the social world—as if such an account were indeed possible. But the search for truth as representation involves neither objectivism, nor a poetic evocation of "nothing."

The current repudiation of realism, and the classical theory of truth as representation by postmodernist anthropologists, seems to be largely due to the baneful influence of Martin Heidegger on the social sciences. The writings of this reactionary thinker now have the status of sacred texts, which are the subject of much exegesis by scores of admiring acolytes (Dallmayr 1993, Foltz 1995). It is reminiscent of the scholastic musings of the early structural Marxists who pored over the writings of Marx in a similar fashion. Along with Nietzsche, Heidegger is seen as a key figure in the rise of postmodernism, though both scholars would probably repudiate entirely the idealism and relativism of their purported followers. Heidegger, in particular, was a realist. In his essential thoughts he was pre- (not post-) modern—something of a peasant philosopher. Most of his key ideas and concepts are, in fact, a recycling of Aristotle's philosophy. As a realist, Heidegger was seeking a primordial or poetic relationship with things (beings) in the world, and he never denied the classical theory of truth as correspondence but rather sought to express—often in mind-boggling abstractions—a primordial sense of truth. This was truth as *aletheia* (derived from a reading of Aristotle), truth as disclosure, as revelation or "unconcealment," which he saw as prior to truth, as the "agreement of intellect and the thing." Thus the phenomenological approach to both nature and history which he advocates—an approach derived from Husserl—promises, he writes, "to disclose reality precisely as it shows itself before scientific inquiry, as the reality which is already given to it" (1985, 2). Heidegger, as his student and friend Hannah Arendt insists (1978, 19), used the terms "meaning" and "truth" loosely and interchangeably. Thus what Heidegger—like Husserl—seems to be suggesting is a quite uncontentious notion, namely that before one can ascertain the truth (as correspondence) or explain a particular phenomenon, one must first disclose its meaning, or ascertain a "primordial" understanding (or, for Husserl, a universal inter-subjective knowledge) of it as a phenomenon.

However, strongly influenced by Nietzsche and German reactionary nationalism, Heidegger had nothing but disclaim for the life-world (*dasein*, social existence) of ordinary people. This life-world was seen as inauthentic, a world of drab uniformity and conformism. The essential message of his famous *Being and Time* (1962) is that only an elite few could escape the vulgarity of the inauthentic existence, and the nihilism, that constitutes contemporary social life (Herf 1954, 111; Wolin 1990, 49–55). Highly critical of capitalism, as befits a peasant reactionary (though capitalism is never mentioned, but alluded to in terms of the "technological mastery" of nature), and of science, Heidegger tends to make a series of conflations that leads one to assume that he is completely repudiating both truth as correspondence (representation) and empirical science.

Heidegger makes an initial distinction between the world "present-at-hand" (*vorhanden*), which seems to imply the intuitive recognition of things in the world, from things "ready-to-hand" (*zuhanden*). Unlike anthropologists such as Scott Atran (1990), Heidegger emphasises that our practical relationship with the world—our "productive knowledge" (*techne*) in Aristotle's terms—has primacy over the "present-at-hand." This latter existential attitude to the world is then linked both to the correspondence theory of truth and to the Greek *theoria*, the rational contemplation of nature. These, in turn, seem to be conflated with scientific realism, which is seen by Heidegger as a "challenging" project, one that involves a confrontation with or opposition to the world. And all these are linked to "metaphysics" which comprises all of Western philosophy since Plato—including Platonic idealism, Cartesian rationalism, positivism, and Husserl's phenomenology. Metaphysics, he tells us, "thinks beings as beings in the manner of representational thinking that gives grounds"—the "ground" showing itself as "presence," expressed in terms of *eidos, idea, arche, ego cogito,* or Nietzsche's will to power. All metaphysics, he tells us, including positivism, "speaks the language of Plato" (Krell 1978, 432–44). Both metaphysics and science are then linked to the "technological mastery" of nature. Heidegger thus interprets the whole of Western philosophy and thought—prior to his good self—in the most obfuscating, monolithic fashion as "metaphysics"—which is then dismissed with some presumption as "onto-theology" (a variant of Platonism).

Truth, for Heidegger, thus does not imply *orthotes*, "representation" (as this concept is used by ordinary mortals who have never heard of Plato)—for this implies ahistoric or absolute truths, or the mastery of nature (given his conflations)—but rather suggests instead *aletheia*, the disclosure of truth. This is a form of poetic thinking, a "clearing of being," a primordial approach to the world that lets the things themselves speak to us (Krell 1978, 442). The slogan "to the things themselves," associated with Hegel's and Husserl's phenomenology, is a clarion call for Heidegger, heralding a new way of "thinking." It is, therefore, not surprising that he is now considered a patron-saint of deep ecologists, although it is well to remember that in his "Letter on Humanism" he critiques Aristotle and the existentialists for even suggesting that humans are rational animals, insisting that only humans have "ek-sistent," and that there is an "abyss" separating humans and animals (Krell 1978, 227–30). Nothing could be more un-ecological than this kind of anthropocentrism. Heidegger never seems to have heard of Darwin, and his interests were very narrowly focused—on Greek and German philosophy. His discussion of the abstruse "fourfold"—mortals, sky, earth, and the divinities (as the messengers of the godhead) (351)—indicate that he remained essentially a theological thinker throughout his life. His philosophy, like everybody else's—was also metaphysical and "grounded" though he never did get around to telling us the meaning of Being, which, he continually tells us, was "concealed" or "forgotten" by every philosopher over the past two millennia. It is important not to get over-awed or mesmerised by Heidegger's philosophical abstractions, and to recognise that his important emphasis on existential hermeneutics (derived from Dilthey, developed by his student Gadamer, and taken for granted by most social scientists) does not imply a repudiation either of *orthotes*, truth as representation, or of empirical science. Nor is it helpful to conflate empirical science with capitalism (power) and with technological reason—although this is not to deny the intrinsic relationship that currently pertains between capitalism and empirical science.

Heidegger largely seems to be insisting—in endless repetition (1962, 257–73; 1994; Krell 1978, 115–38)—what has largely been taken for granted by generations of social scientists, at least those who have distanced themselves from crude positivism, namely, that one must engage in interpretative understanding or hermeneutics (meaning, verstehen) before one can explicate social phenomena through causal analysis or historical reason.

Many contemporary anthropologists, uncritically following Heidegger, and his philosophical acolytes such as Rorty, have the mistaken idea that truth as representation implies either an absolutist metaphysic, the arrogant presumption that what is being portrayed are transcendental verities, or that it implies a Promethean perspective, the control and domination of Nature (see Tyler for this version of truth). Empirical knowledge and truth as representation entail neither of these notions. What knowledge as representation does, however, is to make explicit what in fact is being affirmed (truths about the world), and acknowledge that all truth is inter-subjective and thus open to critical scrutiny and possible refutation by other scholars (unlike truths which are apparently disclosed through evocation or mystical "revelation" and which we are told have no reference at all to any world outside the text). With regard to anthropology, this affirmation of truth as representation is particularly important, for ethnographic accounts and anthropological theory should be open to scrutiny by the people whose culture and social life is being described and explicated. All knowledge is inter-subjective, as even positivists long ago recognised, and thus only approximates to the truth (Feigl 1953). The notion that an earlier generation of anthropologists were enwrapt in a "visualist" metaphor, lacked any imagination, and saw themselves as detached observers, neither participating in the social life of a community nor being engaged in dialogue, is something of a caricature. Such a crude positivistic or behaviouristic stance would never have yielded the rich ethnographies of scholars like Malinowski, Fortes, Evans-Pritchard, Audrey Richards, and Turner, to name but a few.

It is important to recognise that all knowledge, indeed all philosophy, is based on certain metaphysical assumptions. For common sense and science the crucial assumption is that some entities and relations in the world are necessary and relatively enduring—whether one considers the moon, elephants, hunting activities, or cultural traditions. This does not deny that the world is also in flux and is constantly changing, as dialectical (process) philosophy, from Heraclitus to Hegel and Marx, has always recognised. But the postmodernist emphasis on contingency, indeterminacy, and fragmentation, which long ago reached its apotheosis in Hume's skepticism, undermines the whole notion of objective knowledge. Unlike the postmodernists, however, Hume realised that his idealist theories made no sense outside of his study.

Human Agency

Human life is an essential "paradox" (as Husserl describes it), for there is an inherent dualism in social existence, in that we are contemplative beings—and through our conscious experience see ourselves as separate from the world—while at the same time being active participants in this world. But our conscious awareness of the surrounding life-world (*lebenswelt*) essentially expressed in visual metaphors, does not in the least imply that vision, insight and observation entails detachment or distance from the world. Nevertheless, there is, as Aristotle and Kant long ago explored, both a transcendental (theoretical, reflective) and an empirical (practical, productive) dimension to human life. We thus have, in a sense, a dual existence, in that we are simultaneously both contemplative and active beings, both "constituting" (giving meaning to) and being causally related to the world. Hence the paradox: of humans as "world-constituting subjectivity" as well as being objectively and actively incorporated in the world (Husserl 1970, 262). In similar fashion, we are both personal and social beings, ontologically distinct from social life but at the same time constituted through it.

Social reality, unlike the material world, is not self-subsistent but is dependent on human activity. At the same time, like nature, it is transformed and changed by human agency. There is then an "ambivalence," as Margaret Archer (1995, 2) writes, about social reality, in that social and cultural structures are dependent upon human activity, while at the same time, through social praxis, these structures have a constraining and determining influence over us. As Karl Marx long ago put it: "Men make their own history, but they do not make it just as they please; they do not make it under circumstances chosen by themselves, but under circumstances directly encountered, given and transmitted from the past. The tradition of all the dead generations weigh like a nightmare on the brain of the living" (Marx and Engels 1968, 96). Human beings thus, through social praxis, create social and normative structures, and cultural schemas, which in turn, as emergent entities, constrain and condition human consciousness and behaviour. The history of the social sciences has thus long been an ongoing debate—bordering on a dispute—between two quite distinct ontologies or approaches to social life. Indeed Alan Dawe (1979) has interpreted the "persistent tension" between the two approaches as an immediate expression of the inherent "dualism of social experience." The first approach has been described variously as holism, collectivism or "social systems" theory; the second approach as atomism, methodological individualism, or "social action" theory (Cohen 1968; Hollis 1994, 5–12; Archer 1995, 34–54).

The first approach, holism, puts a decisive emphasis on "society," on social structures as the fundamental reality, and treats human agency and consciousness as epiphenomenal. Durkheim's sociology—along with that of Talcott-Parsons—and Marx's famous preface to *A Contribution to the Critique of Political Economy* (1859), are taken as exemplifying this approach. In a famous phrase Marx wrote: "It is not the consciousness of men that determines their being, but, on the contrary, their social being that determines their consciousness" (Marx and Engels 1968, 181). This "top down" approach, or what Archer describes as "downward conflation," which presents humans as "Homo Sociologicus" and dissolves personal identity and human agency into social structures, is characteristic of normative functionalism (Talcott-Parsons), mechanistic versions of Marxism, much symbolic anthropology, as well as of structuralism. This approach has indeed, as we have noted earlier, been given a new lease of life by the postmodernists, who in a critique of the transcendental subject of Cartesian metaphysics, have gone to extreme in eradicating human agency from the analysis. Thus the human person has either been "erased" entirely—the "end of man" syndrome—and human history seen as a "process without a subject" (Althusser) or the self is seen as simply an "effect" or "construction" of ideological discourses or language (see Coward and Ellis 1977, 74; Flax 1990, 231).

Although acknowledging the reality of social structures and cultural representations, and their constraining influence, via social praxis, on human life, it is equally important to affirm the salience of human identity and agency. But the human person must be conceptualised not as a transcendental epistemic subject but as an embodied being, embedded in both an ecological and a social context (Benton 1993, 103; Morris 1994, 10–15). Humanism, as a belief in the power of human agency in history, must therefore be affirmed and should not be equated either with bourgeois individualism (reflected in political liberalism, economic theory, or rational choice strategy), nor with Cartesian metaphysics, nor with the notion that this necessarily implies a technological mastery of nature. Further, humanism, the emphasis on human agency, does not entail the deification of either reason or humanity, as Ehrenfeld, in his study *The Arrogance of Humanism* misleadingly suggests in his disguised defence of religion (see Bookchin 1995, 12–14; Assiter 1996, 75).

The second approach, that of individualism, rests upon the empiricist conviction that the ultimate constituents of social reality are human individuals or people, and thus treats social structures and cultures as epiphenomena. The prototype statement comes from John Stuart Mill who wrote: "The laws of the phenomena of society are, and can be, nothing but the laws of the actions and passions of human beings united together in the social state" (1843, Book VI, Chapter 7). This became the fundamental stance of many methodological individualists and empiricists who affirm that the "ultimate constituents of the social world are individual people who act more or less appropriately in the light of their dispositions and understanding of their situation" (Watkins 1968, 270). Marx and Engels are also linked to this approach for they wrote (1845): "History does nothing; it does not possess immense riches, it does not fight battles. It is men, real living men, who do all this, who possess things and fight battles. . . .

History is nothing but the activity of men in pursuit of their ends" (1975, 93; Flew 1985, 61–66). As distinct from the "view from on high," individualism or social action theory is a "bottom up" approach, a form of "upwards conflation," reducing social structures and cultural schemas to the social interactions and dispositions of individual humans. This approach has not only been associated with methodological individualists (Popper, Hayek, Homans, Ayn Rand), and the "rational agent" of economic theory (along with rational choice and games theory), but also with various forms of interpretative sociology—the phenomenology of Schultz and R.D. Laing, ethnomethodology and symbolic interactionism (Layder 1994, 58–74; Hollis 1994, 115–41; Archer 1995, 34–46).

There have been many recent attempts to transcend the "duality" of structure and agency, namely Giddens's (1984) structuration theory and his "ontology of praxis," Foucault's (1980) "genealogical" approach and his emphasis on "power-knowledge," Elias's (1978) theorising around the concept of "figuration," and Jackson's (1989) radical empiricism may be taken as examples. But what all these approaches tend to entail is a conflation of the central dialectic between human agency and social structure—a dialectic which is oblated through an undue emphasis on, respectively, social practices, discourses of power, social configurations, or lived experience (see Layder 1994, 94–149).

What many writers have thus suggested is the need for a critical realism that acknowledges both human agency and social structure (or cultural representations) as distinct levels of reality, each having emergent properties that are irreducible to one another. As Margaret Archer writes, in her advocacy of a morphogenetic approach to sociology, social structures as emergent entities: "are not only irreducible to people, they pre-exist them, and people are not puppets of structures because they have their own emergent properties which mean they either reproduce or transform social structure rather then creating it" (1995, 71). Although Archer refers to this approach as "analytic dualism," she advocates a dialectical not a dualistic approach, an analysis that "links" people to their social context, without implying any reduction. The inter-play between structure and agency is historical, in that the "morphogenetic cycle" involves social structures which in a sense are prior to, and condition, social interaction and agency, which, in turn, transforms or reproduces the social structures (1995, 165–94). She repudiates any recourse to metaphor in describing social reality, for society is not like a language or text, it is not a mechanism with fixed parts, nor is it a cybernetic (homeostatic) system, nor yet is it a kind of theatrical performance or a piece of textile fabric (a common metaphor among radical empiricists)—but rather society is only itself, ordered, open, processual, and peopled (166).

Both classical approaches to social life are inadequate. An emphasis on methodological individualism reduces social life to human dispositions and interactions—and humans are invariably depicted as Hobbesian individuals, rational, competitive, self-interested, autonomous, maximising their own utility. This implies a form of voluntarism. On the other hand, in treating collective phenomena—social structures, cultural representations, discourses, or language—as sui genesis, as the fundamental reality, with regard to which humans are simply an "effect," entails a reification of social phenomena and is equally untenable. A human being, as Erich Fromm put it, is not "a blank sheet of paper on which culture can write its text" (1949, 23)—for humans are beings with natural capacities, charged with energy, and structured in specific ways. Humans, and their self-identity, are not simply an "effect" of social structures, or of discourses.

What is therefore needed is a theoretical perspective that combines both approaches. It was a perspective initiated by Marx, who, significantly, is seen as exemplifying both approaches, for indeed, Marx emphasised both human agency and the ontological reality of social institutions and ideologies. What then is needed is an approach that combines both humanism and structuralism, providing a "linkage theory" or a "relational model" of society. This suggests that our social being is constituted through social praxis for our social acts presuppose the existence of social institutions and cultural schema and beliefs. Yet, at the same time, humans are seen as ontologically independent of social relations, and, as personal beings, have agency, self-consciousness and self-identity (Harre 1983). There is then a need—in this era of postmodernism—to reclaim human agency for social analysis—without this implying a relapse into transcendental subjectivity, or the acceptance of the bourgeois individual espoused by economists and rational choice theorists. The emphasis on social structure indicates the way in which cultural schemas (language) and social institutions come, through social praxis, to shape and modify human consciousness and behaviour. The emphasis on human agency highlights the degree to which humans change social structures and cultural frameworks (Collier 1994, 140–41; Layder 1994, 209–10).

Many scholars, of course, throughout the history of the social sciences, have attempted such an analysis. Among more recent scholars who have exemplified this perspective are Habermas and Bourdieu, whose theories have been subject to much recent analysis (Bourdieu 1990, Outhwaite 1994).

In a paper written more than a decade ago (1985)—specifically focused on radical scholars who had attempted to bring a human/psychological dimension into social analysis—Kardiner, Reich, Fromm, and Laing—I critiqued the two extreme positions taken on the siting of human agency within the social sciences. The one extreme was to give human subjectivity and agency absolute priority in social analysis (a strategy assumed by ego psychologists, some psychoanalytic writers, rational choice theorists, transactional analysis and methodological individualism); the other extreme was to expunge human agency from the analysis entirely—as suggested by Leslie White, behaviourists, many semiologists and the structural Marxists (and of course, since then, by many postmodernists). My own thoughts on the "theory of the subject" (i.e., human agency), followed, I suggested, those of Lucien Goldman. Like Marx, Goldman rejects as untenable the many radical dualisms that pervade contemporary thought—philosophy and science, theory and praxis, interpretation and explanation. And on the concept of the subject he argues persuasively against the two main approaches. One gives the subject, specifically the *cogito*, analytical priority—a line of thought that stems from Descartes and is expressed by existentialists, phenomenologists, as well as by humanistic psychologists and interpretive sociologists. The other, characteristic of contemporary structuralism (and he specifically cites Althusser and Lévi-Strauss, whose theories Coward and Ellis closely follow), leads to the "negation of the subject." Although the "subject" has the status of other scientific concepts in that it is constructed, Goldman holds that such a concept is a grounded one in having a necessary function of "rendering the facts we propose to study intelligible and comprehensible" (1977, 92). The first approach, which begins with the individual subject and puts a focal emphasis on meaning, is, Goldman argues, essentially non-explanatory and unable to account for the relationship among phenomena. The second approach, by negating the subject, is unable to account for the becoming or genesis of a structure, or its functioning. "The first does not see structure; the second does not see the subject which creates genesis, becoming and functionality" (1977, 106). Goldman, therefore, argues that it is necessary to integrate consciousness with behaviour and praxis, and to seek both the meaning and functionality of structures. To do this one needs a dialectic approach that situates a "creative subject at the interior of social life." "To comprehend a phenomenon is to describe its structure and to isolate its meaning. To explicate a phenomenon is to explain its genesis on the basis of a developing functionality which begins with a subject. And there is no radical difference between comprehension and explication" (1977, 106). But to argue against the theoretical negation of the human subject is not in the least to deny that the "self" or "mind" is socially constituted; the subject, human existence, is no more conceivable outside social relations, than are social relations conceivable without subjects (Morris 1985, 736).

It should be recognised, however, that the ontological distinction between social structures and human agency, and the respective emphasis given by the methodological individualists and "collectivist" scholars—the latter stressing the priority of social structures, cultures, or language—is not coterminous with the epistemological division between hermeneutics and naturalism. For example, Mill and Hayek on the one hand and Radcliffe-Brown and Leslie White on the other, all advocate a scientific approach to social life, and tend to play down hermeneutics, yet they are on different sides of the fence in terms of their metaphysics. Indeed, White suggested that the most adequate scientific interpretations of culture proceeded as "if human beings did not exist" (1949, 141). Likewise, Durkheimian sociology, structural Marxism, and postmodernist anthropology and philosophy have much in common in their repudiation of human agency; but postmodernism puts a crucial emphasis on an extreme form of hermeneutics—deconstruction or textualism—and disavows both science and empirical knowledge. It is to this latter issue that I may now turn.

Empirical Knowledge

Reflecting on the later writings of Evans-Pritchard in 1961 David Pollock suggested that his mentor had instigated a movement in anthropology from "function to meaning" (1961, 76). This thesis completely overlooks the fact that one of the doyens of functionalist anthropology, Radcliffe-Brown, had long emphasised the importance of delineating the "meaning" of social phenomena, and that Malinowski—the other founding father of British social anthropology (Kuper 1973)—had also stressed the importance of understanding the "native's point of view" (Radcliffe-Brown 1922, ix; Malinowski 1922, 25). Cultural anthropology in the United States, long before Geertz, had also emphasised the crucial importance of interpretive understanding, as the work of Cushing, Boas, Reichard, and Benedict attest. It is also significant to note that Evans-Pritchard's important critique of Radcliffe-Brown's positivism, rather than advocating a hermeneutic or symbolic approach to anthropology, stressed the need to create a dialogue between anthropology and historical understanding (Evans-Pritchard 1962; Morris 1987, 188–89).

In an important sense, then, hermeneutics and interpretive understanding has always been a constituent part of anthropology, especially in relation to ethnographic studies. It has been accepted as such even by cultural materialists like Marvin Harris. What Harris challenged in the cultural idealist tradition of Geertz and Schneider was not their descriptive hermeneutics or "cognitivism" per se, but rather the tendency to reduce all social life to semiotics and to repudiate causal analysis and historical understanding, to deny, that is, that cultural representations are explicable in terms of "infrastructural conditions" (Harris 1980, 258–82).

The so-called "interpretive turn" in the social sciences, which is reputed to have occurred some thirty years after Evans-Pritchard's seminal analysis of *Nuer Religion* (1956), has to be understood largely as a *reaction* to positivistic sociology and to the scientism of much structural Marxism and structuralist analysis. Indeed, Josef Bleicher, writing in the early 1980s had suggested that with the radical influence of positivistic science, that there had been the concomitant "atrophy" or "demise" of the "hermeneutic imagination" within the human sciences (1982, 1–2). A decade later the pendulum, it seems, had swung to the other extreme, and Rabinow and Sullivan's (1987) advocacy of the "interpretive turn" in the social sciences— along with postmodernism—seem to many to entail the complete repudiation of empirical science. But Rabinow and Sullivan made it clear that the interpretive approach, with its focus on "cultural meaning" and with the emphasis that knowledge as "practical" and historically situated, did not imply a lapse into cultural relativism, or the exaltation of a romantic "subjectivism." And they were dismissive of Derrida's "textualism," in that it completely oblates social praxis. But as with Geertz, their fundamental emphasis is on anthropology as a form of hermeneutics, involving the "interpretation of culture," the latter defined as a "web of signification," as the shared meanings practices and symbols that constitute the human world (1987, 7). They do not deny the persistence and theoretical fruitfulness of certain "explanatory schemas" in the social sciences but these are never theorised. The aim of social science, they write, is not to uncover "universal laws" but rather to "explicate" contexts. What such explication involves however is again not specified or explained—given their stress on "hermeneutics" and the interpretation of cultural meanings.

A similar standpoint is taken by Charles Taylor (1985) who in his advocacy of a "philosophical anthropology" stresses the crucial importance of a "hermeneutic component" in the human sciences. But what he critiques as "naturalism" is largely an outdated conception of empirical science,. which involves: a mechanistic paradigm, a "designative account" of meaning which implies an unmediated relationship between language and the world, and a "disengaged identity," a conception of the person that is disembodied and atomistic, divorced from the social context. Such a "natural science" (i.e., positivist) model has long been critiqued by the social sciences, without this entailing the repudiation of naturalism, empirical science. Taylor, however, seems to have little sympathy for those dismissive of the scientific outlook, and even less sympathy towards the obscurity and posturing that is reflected in Derrida's writings. Yet although arguing that interpretation is essential to explanation in the human sciences, Taylor, like Rabinow and Sullivan, is silent when it comes to exploring what "explanation" exactly entails.

One can surely recognise that humans are an intrinsic part of nature, and that social life is explicable by means of empirical science—naturalism—without this implying a reductive, mechanistic and atomistic conception of science (positivism). Likewise, one can recognise the fundamental importance of hermeneutics in the social sciences, without this implying "textualism" and the uncritical embrace of postmodernism. What then is surely needed is an approach that combines science (naturalism) and hermeneutics (humanism), avoiding the extremes of both positivism, which repudiates hermeneutics and tends towards reductive materialism, and textualism, which repudiates empirical science and tends towards cultural idealism.

Hermeneutics, as interpretive understanding, has always been intrinsic to anthropology, especially to the fieldwork experience. Indeed, as Bleicher writes, the social sciences contain a "hermeneutic dimension which is both ineradicable and foundational" (1982, 2). For "meaning" is a central category in the study of social phenomena. A purely positivistic approach to social life, one that tends to repudiate hermeneutics (see Abel 1948, B.F. Skinner 1953), is therefore untenable. Positivism has been the subject of numerous critiques within the social sciences, critiques that long predate the so-called literary or interpretive turn in anthropology. Positivism, or logical empiricism, has been defined by the following criteria:

1. It takes an empirical conception of knowledge, which holds that all knowledge is based on sense experience. Science is thus held to deal with observable "facts" that are completely independent of the researcher.
2. Science involves the search for general laws, in which events or phenomena are explained by reference to universal generalisations, causal regularities based on "constant conjunction."
3. Science is the only form of valid knowledge, all other modes of understanding—poetry, morality, practical knowledge, art, religion, philosophy—are dubbed "metaphysics" and seen as "meaningless," or as offering us no cognitive understanding of the world. An absolute dichotomy is thus maintained between facts and values, and philosophy is seen simply as the "handmaid" of science.
4. Scientific knowledge is viewed as having instrumental value, giving us control over the material world.
5. Scientific rationality is a unified form of knowledge and its method can be applied to both the social sciences and the humanities—indeed, this form of knowledge has no recognisable limits. As Carnap put it: "there is no question whose answer is in principle unattainable by science" (1967, 290). The unity of the sciences tends to give epistemic privilege to mathematics and physics and to entail a form of reductive naturalism (Outhwaite 1987, 5–11; Sorell 1991, 3–22; Hollis 1994, 40–65).

One can, of course, acknowledge the importance of science as a creative and imaginative representation of reality, one whose accuracy is tested by various—and different—practices of validation, without accepting either the positivistic conception of science, or its deification. Neither Aristotle nor Kant accepted that science was the only form of knowledge, and a realist conception of science has long been articulated. Acknowledging that science is a social practice does not deny the validity of empirical science; equally the fact that scientific theories and hypotheses are always "mediated" by social praxis does not entail a repudiation of the correspondence theory of truth (Reyna 1994).

It is of interest that none of the major scholars whose theories are invoked to support the alleged "interpretive turn" in the social sciences or the hermeneutic critique of positivism—Dilthey, Gadamer, and Ricoeur—ever disavowed empirical science. To the contrary, they acknowledged the importance of both hermeneutics and empirical science.

Dilthey's *lebensphilosophie* was a reaction both against excessive rationalism and the extension of positivism to the human sciences. What he attempted was to lay the foundations of a "historical science," one that was based on "lived experience" (*erlebnis*) and acknowledged that the human scientist was also in essence a historical being. But, for Dilthey, empirical knowledge went beyond hermeneutic understanding, for such understanding needed to be complemented by other procedures, especially the identification of causal regularities. For Dilthey, as Outhwaite writes, "did not believe that the use of *verstehen* (interpretive understanding) ruled out causal explanation based on comparison and generalisation; the two methods are complementary," and both were used in the historical (cultural) sciences (1975, 29). Dilthey—like Marx and Weber—thus rejected the dualism between understanding and explanation, for the historian or sociologist would draw on causal explanation when referring to the influence of the social milieu. Weber expressed Dilthey's position well when he defined sociology as "a science which attempts the interpretive understanding of social action in order thereby to arrive at a causal explanation of its cause and effects" (1947, 88) (For useful accounts of Dilthey see Makkreel 1975; Outhwaite 1975, 24–37; Bleicher 1982, 55–68; Morris 1991, 143–52).

While Dilthey saw interpretive understanding as a constituent part of the cultural sciences, Gadamer, who was critical of Dilthey's attempt to make hermeneutics into a science, tended to view science as a form of interpretation. But for Gadamer, the attempt on the part of the human sciences to emulate the physical sciences could only lead to a kind of "intellectual sclerosis," and to underplay the crucial importance of "meaning" with regard to social phenomena (Bleicher 1982, 88). Gadamer, however, was less concerned to attack science than to emphasise the "historicity" of human subjectivity and the fundamental importance of cultural traditions, and thus to reaffirm the importance of history, the arts, and other forms of human knowledge. Gadamer argues that historical understanding does not imply "subjectivism," but rather "the placing of oneself within a tradition, in which past and present are constantly fused." "Understanding" was therefore not simply a "method" of the social sciences, but rather it was the "original characteristic of the being of human life itself." Thus the natural science mode of knowledge is viewed as a "subspecies of understanding" (Gadamer 1975, 258–59). For Gadamer, however, hermeneutics did not entail either cultural relativism or idealism; it involved what he described as a "fusion of horizons," the wider expansion of human knowledge (302–6). In emphasising the historical nature of our beliefs and practices, Gadamer does not suggest that "we must give up a concern with reason, with the validity of our knowledge, but rather one must preserve the Enlightenment ideal, while rendering it compatible with the cultural and linguistic embeddedness of our understanding" (Warnke 1987, 168).

In a similar fashion, while Paul Ricoeur emphasises the crucial importance of hermeneutics in the social sciences he does not repudiate—any more than does Dilthey and Gadamer—the importance of empirical science as a mode of explanatory knowledge. Deeply rooted in the phenomenological tradition, Ricoeur follows Husserl and Heidegger in affirming the ontological priority of the life-world (*lebenswelt*), such that hermeneutics (existential understanding) precedes both interpretation (as exegesis) and the explanation of social phenomena. As he writes, "Hermeneutics is not a reflection on the human sciences but an explication of the ontological ground upon which these sciences can be constituted." But an emphasis only on the primordial understanding of the life-world, as with Heidegger, Ricoeur finds extremely limiting. With Heidegger's philosophy, he writes, "we are always engaged in going back to the foundations, but we are left incapable of beginning the movement of return which would lead from the fundamental ontology to the properly epistemological question of the status of the human sciences. Now a philosophy which breaks the dialogue with the sciences is no longer addressed to anything but itself" (1981, 59). Although recognising the inherent polysemic nature of language, and critical of the notion that language is a "mirror of nature" (logocentrism), Ricoeur is equally critical of Derrida's deconstructionist approach, in that the emphasis on absolute indeterminancy between language and the world invalidates all claims to a sense of order—indeed to any sense of meaning (see Rabinow and Sullivan 1987, 13). Only a dialectic of sense and reference, Ricoeur suggests, can tell us something about the relationship between language and the ontological condition of being in the world (1976, 20–21). But though emphasising, like Dilthey and Gadamer, the ontological importance of lived experience (*erlebnis*), the fundamental "historicity" of human life, and hermeneutics as a "primary experience," Ricoeur does not disavow the "explanatory attitude." He seems to recognise three "movements" in the knowledge that is conveyed by the human sciences—a prior understanding (*verstehen*) that involves becoming a participant in a human life-world (*lebenswelt*), the interpretation (*auslegung*), a reflective exegesis of this life-world (culture), and finally explanation (*erklären*). But explanation, for Ricoeur arises from within the human scientific tradition itself and does not imply the imposition of a positivistic model of science onto social and cultural phenomena. He thus claims that meaningful action may become an object of science, in the same way that writing is a kind on objectification, the understanding of a text entailing a movement "from sense to reference" (1981, 218). He therefore advocates a complementarity and reciprocity between interpretation and explanation, between hermeneutics and the kind of structural analysis suggested by Lévi-Strauss (150–53).

All these three scholars stress the importance of the hermeneutic imagination in the human sciences—Dilthey, Gadamer, and Ricoeur—and thus in essence repudiate the stark neo-Kantian dualism between humanism (hermeneutics) and naturalism (explanation). Thus in reviewing the hermeneutic tradition, Bleicher concludes that the importance of hermeneutics does not preclude a recognition of the contextuality and historicity of social phenomena, and that both interpretation and explanatory accounts (science) draw on the capacity to understand. Bleicher therefore advocates a sociology that involves a combined "hermeneutic—dialectical" analysis, one that is critical, hermeneutically informed, as well as cognisant of the objective contexts in which communicative processes occur (1982, 143–50).

Anthropology, it seems to me, has always tried to maintain a bridge between the natural sciences and the humanities. Long before postmodernism, there have been those who have denied that the discipline is, or ever can be, a science. Equally, there have been those, like Radcliffe-Brown, who have been adamant that anthropology should be modelled on the natural sciences and either be concerned through comparative studies to establish causal laws or inductive generalisations, or, in the fashion of the structuralists, to delineate universal structures. Most anthropologists, however, especially those with a sense of history, have tended to occupy the middle ground. Following a long tradition that begins with Vico, they have thus been concerned with both interpretive understanding and scientific explanation, without collapsing into a crude positivism. They have thus tried, in various ways, to unite the Enlightenment and romantic traditions.

In a critique of the over-emphasis on interpretation, Tim O'Meara (1989) has argued that explanations of human behaviour have always been an intrinsic part of anthropology, and that anthropology is thus an empirical science. The notion that anthropology is simply a romantic rebellion against the Enlightenment, an approach that disdains any possibility of an empirical science with relation to social phenomena, as suggested by Richard Shweder (1984), is, I think, quite misleading. It is simply an early expression of postmodernist anthropology (see my critique, Morris 1986).

Human life is inherently social and meaningful, as well as being "enmeshed" or "rooted" in the natural world. An understanding of human life therefore entails both hermeneutic understanding and interpretation (humanism) as well as explanations in terms of causal mechanisms and historical understanding (naturalism). Anthropological analysis must combine both hermeneutics and naturalism, and avoid the one-sided emphasis either on hermeneutics—which in its extreme form, "textualism," denies any empirical science—or on naturalism—which in its extreme form, "positivism," oblates or downplays cultural meanings and human values. Thus, as Jeffrey Alexander argues (1992), the "epistemological dilemma" between positivism (scientific theory) and hermeneutics (anti-theoretical relativism) is a false dichotomy which needs to be rejected. The alternative to positivistic science and objectivism is not a facile acceptance of a neo-romantic textualism or hermeneutics, that espouses an idealistic metaphysic and cultural relativism. Thus anthropology must continue to follow the tradition of the historical sociologists (Marx, Dilthey, Weber, Evans-Pritchard) and in combining hermeneutics (interpretive understanding) and empirical science (explanations) repudiate both textualism and positivism. As Jackson writes: "People cannot be reduced to texts any more than they can be reduced to objects" (1989, 184). But in acknowledging a naturalistic perspective, and the psychic, moral and epistemological unity of humankind, this does not imply the "destruction" of the concrete, the cultural particular, or the historical, nor does it imply that people's behaviour is the same everywhere—as Hollinger seems to believe (1994, 67). Unity, difference, and singularity are all dimensions of the world, and of human life.

We can thus but conclude by defending anthropology as an empirical science, and following Bhaskar, suggest that it is concrete (in the sense of Husserl), hermeneutical (in the sense of Dilthey) and historical (in the sense of Marx), and that social phenomena can therefore be understood through interpretive understanding, causal analysis and historical reason (Bhaskar 1986, 186; Collier 1994, 163–70).

Anthropological knowledge therefore consists of different kinds of knowledge—the prior, practical understanding of a given life-world or

culture, interpretative analysis that reflectively attempts to disclose underlying forms or structures of a given society or culture, and the delineation
of causal mechanisms through comparative studies, and by situating the
phenomena, by means of the historical imagination, in a socio-historical
context (Evans-Pritchard 1962, 23; Comaroff 1992). This does not entail
a simple and stark division of labour between ethnography (interpretation) and anthropology. Dan Sperber suggests (1985, 34) that the task of
ethnography is interpretation, making intelligible the experience of a
particular social domain, and that the task of anthropology as a gener-
alising science is to describe the causal mechanisms that account for a
cultural phenomenon. He sees these as two autonomous tasks, which are
best initially divorced from each other. But ethnographic texts themselves
often incorporate different research strategies, and thus include historical
and comparative studies and causal explanations, as well as hermeneutic
understanding. But an anthropology worthy of the name seems to me
to embrace what Adam Kuper (1994) has described as the "cosmopolitan
project"—linking with other social sciences (as well as with the humanities) to contribute towards the comparative understanding of human life.
It thus affirms a realist ontology and acknowledges that anthropological
understanding is a search for truth. It also affirms the salience of human
subjectivity and combines hermeneutics (the descriptive understanding
of social life) with the advocacy of an empirical social science.

References

Abel, Theodore. 1948. "The Operation Called Verstehen." In *Readings in the Philosophy of Science*,
    edited by Herbert Feigl and May Brodbeck, 677–87. New York: Appleton-Century.
Alexander, Jeffrey. 1992. "General Theory in Post-positivist Mode." In *Postmodernism and Social
    Theory*, edited by Steven Seidman and David G. Wagner, 322–68. Oxford: Blackwell.
Archer, Margaret S. 1988. *Culture and Agency: The Place of Culture in Social Theory*. Cambridge:
    Cambridge University Press.
_____. 1995. *Realist Social Theory: The Morphogenetic Approach*. Cambridge: Cambridge
    University Press.
Arendt, Hannah. 1978. *The Life of the Mind*. New York: Harcourt Brace.
Aristotle. 1925. *The Nicomachean Ethics*. Translated by D. Ross. New York: Oxford University
    Press.
Assiter, Alison. 1996. *Enlightened Women*. London: Routledge.
Atran, Scott. 1990. *Cognitive Foundations of Natural History*. Cambridge: Cambridge University
    Press.
Benedict, Ruth. 1934. *Patterns of Culture*. London: Routledge & Kegan Paul.
Benhabib, Seyla. 1995. "Feminism and Postmodernism: An Uneasy Alliance." In *Feminist
    Contentions*, edited by Seyla Benhabib et al., 17–34. London: Routledge.
Benton, Ted. 1993. *Natural Relations: Ecology, Animal Rights, and Social Justice*. London: Verso.
Ben-Ze'ev, Aaron. 1995. "Is There a Problem in Explaining Cognitive Progress?" In *Rethinking
    Knowledge: Reflections across the Disciplines*, edited by Robert F. Goodman and Walter R.
    Fisher, 41–56. Albany: State University of New York Press.

Bhaskar, Roy. 1978. *A Realist Theory of Science*. Sussex: Harvester Press.

————. 1986. *Scientific Realism and Human Emancipation*. London: Verso.

————. 1989. *Reclaiming Reality*. London: Verso.

Bleicher, Josef. 1982. *The Hermeneutic Imagination: Outline of a Positive Critique of Scientism and Sociology*. London: Routledge & Kegan Paul.

Bookchin, Murray. 1990. *The Philosophy of Social Ecology*. Montreal: Black Rose Books.

————. 1995. *Re-enchanting Humanity*. London: Cassell.

Bourdieu, Pierre. 1990. *The Logic of Practice*. Cambridge: Polity Press.

Callinicos, Alex. 1989. *Against Postmodernism*. Cambridge: Polity Press.

Caplan, Patricia, ed. 1987. *The Cultural Construction of Sexuality*. London: Tavistock.

Carnap, Rudolf. 1967. *The Logical Structure of the World*. Translated by Rolf A. George (Berkeley: University of California Press).

Carrithers, Michael. 1992. *Why Humans Have Cultures*. New York: Oxford University Press.

Cassirer, Ernst. 1951. *The Philosophy of the Enlightenment*. Princeton, NJ: Princeton University Press.

Clifford, James, and George Marcus 1986. *Writing Culture*. Berkeley: University of California Press.

Cliteur, Paul. 1995. "The Challenge of Postmodernism to Humanism." *New Humanist* 10, no. 3: 4–9.

Cohen, Percy S. 1968. *Modern Social Theory*. London: Heinemann.

Collier, Andrew. 1994. *Critical Realism: An Introduction to Roy Bhaskar's Philosophy*. London: Verso.

Comaroff, John, and Jean Comaroff. 1992. *Ethnography and the Historical Imagination*. Boulder: Westview Press.

Coward, Rosalind, and John Ellis. 1977. *Language and Materialism*. London: Routledge & Kegan Paul.

Dallmayr, Fred R. 1993. *The Other Heidegger*. Ithaca: Cornell University Press.

Danto, Arthur C. 1968. *What Philosophy Is*. New York: Harper & Row.

Davidson, Donald. 1984. *Inquiries into Truth and Interpretation*. Oxford: Clarendon Press.

Dawe, Alan. 1979. "Theories of Social Action." In *A History of Sociological Analysis*, edited by Tom Bottomore and Robert Nisbet, 362–417. London: Heinemann.

Derrida, Jacques. 1976. *Of Grammatology*. Translated by G.C. Spivak. Baltimore: Johns Hopkins University Press.

Devitt, Michael. 1984. *Realism and Truth*. Oxford: Blackwell.

Elias, Norbert. 1978. *What Is Sociology?* London: Hutchinson.

Evans-Pritchard, E.E. 1956. *Nuer Religion*. Oxford: Clarendon Press.

————. 1962. *Essays in Social Anthropology*. London: Faber and Faber.

Fabian, Johannes. 1994. "Ethnographic Objectivity Revisited." In *Rethinking Objectivity*, edited by Allan Megill. Durham: Duke University Press.

Feigl, Herbert. 1953. "The Scientific Outlook: Naturalism and Humanism." In *Readings in the Philosophy of Science*, edited by Herbert Feigl and May Brodbeck, 8–18. New York: Appleton-Century-Crofts.

Flax, Jane. 1990. *Thinking Fragments: Psychoanalysis, Feminism, and Postmodernism in the Contemporary West*. Berkeley: University of California Press.

————. 1995. "Responsibility without Grounds." In *Rethinking Knowledge: Reflections across the Disciplines*, edited by Robert F. Goodman and Walter R. Fisher, 147–67. Albany: State University of New York Press.

Flew, Antony. 1985. *Thinking about Social Thinking*. London: Fontana.

Foltz, Bruce V. 1995. *Inhabiting the Earth: Heidegger, Environmental Ethics, and the Metaphysics of Nature*. New Jersey: Humanities Press.

Foucault, Michel. 1980. *Power/Knowledge*. Brighton: Harvester Press.

Fromm, Erich. 1949. *Man for Himself*. London: Routledge & Kegan Paul.

Gadamer, Hans-Georg. 1975. *Truth and Method*. London: Sheed & Ward.

Gare, Arran. 1993. *Nihilism Incorporated*. Burgendare: Eco-Logical Press.

Gellner, Ernest. 1995. *Anthropology and Politics: Revolutions in the Sacred Grove*. Oxford: Blackwell.

Giddens, Anthony. 1984. *The Constitution of Society: Outline of the Theory of Structuration*. Cambridge: Polity Press.

Goldman, Lucien. 1977. *Cultural Creation in Modern Society*. Oxford: Blackwell.

Goody, Jack. 1995. *The Expansive Moment: The Rise of Social Anthropology in Britain and Africa, 1918–1970*. Cambridge University Press.

Harré, Rom. 1983. *Personal Being: A Theory for Individual Psychology*. Oxford: Blackwell.

Harris, Marvin. 1980. *Cultural Materialism: The Struggle for a Science of Culture*. New York: Random House.

Hastrup, Kirsten. 1995. *A Passage to Anthropology: Between Experience and Theory*. London: Routledge.

Heidegger, Martin. 1962. *Being and Time*. Oxford: Blackwell.

_____. 1994. *Basic Questions of Philosophy*. Bloomington: Indiana University Press.

Herf, Jeffrey. 1984. *Reactionary Modernism: Technology, Culture, and Politics in Weimar and the Third Reich*. Cambridge: Cambridge University Press.

Hindess, Barry, and Paul Q. Hirst 1975. *Pre-Capitalist Modes of Production*. London: Routledge & Kegan Paul.

Hollinger, Robert. 1994. *Postmodernism and the Social Sciences: A Thematic Approach*. London: Sage.

Hollis, Martin. 1994. *The Philosophy of Social Science: An Introduction*. Cambridge: Cambridge University Press.

Husserl, Edmund. 1970. *The Crisis of European Sciences and Transcendental Phenomenology*. Evanston: Northwest University Press.

Jackson, Michael. 1989. *Paths Towards a Clearing: Radical Empiricism and Ethnographic Inquiry*. Bloomington: Indiana University Press.

Johnson, Mark. 1987. *The Body in the Mind: The Bodily Basis of Meaning, Imagination, and Reason*. Chicago: University of Chicago Press.

Kant, Immanuel. 1934. *Critique of Pure Reason*. London: Dent.

Krell, David F. 1978. *Basic Writings: Martin Heidegger*. London: Routledge.

Kuper, Adam. 1973. *Anthropologists and Anthropology: The British School, 1922–1972*. Harmondsworth: Penguin.

_____. 1994. "Culture, Identity and the Project of Cosmopolitan Anthropology." *Man* 29, no. 3: 537–54

Layder, Derek. 1994. *Understanding Social Theory*. London: Sage.

Makkreel, Rudolf A. 1975. *Dilthey: Philosopher of the Human Sciences*. Princeton, NJ: Princeton University Press.

Malinowski, Bronislaw. 1922. *Argonauts of the Western Pacific*. London: Routledge.

Marx, Karl. 1975. *Early Writings*. Harmondsworth: Penguin.

Marx, Karl, and Friedrich Engels. 1968. *Selected Works*. London: Lawrence & Wishart.

_____. 1975. *The Holy Family*. London: Lawrence & Wishart.

Medawar, Peter. 1982. *Pluto's Republic*. New York: Oxford University Press.

Megill, Allan. 1994. *Rethinking Objectivity*. Durham: Duke University Press.

Mill, John Stuart. 1843. *A System of Logic*. London: J.W. Parker.

Miller, David. 1983. *A Pocket Popper*. London: Fontana.

Moore, Henrietta L. 1994. "Understanding Sex and Gender." In *Companion Encyclopedia of Anthropology*, edited by Tim Ingold, 813–30. London: Routledge.

Morris, Brian. 1985. "The Rise and Fall of the Human Subject." *Man* 20: 722–42.

———. 1986. "Is Anthropology Simply a Romantic Rebellion against the Enlightenment." *Eastern Anthropology* 39, no. 4: 359–64.

———. 1987. *Anthropological Studies of Religion*. Cambridge University Press.

———. 1991. *Western Conceptions of the Individual*. Oxford: Berg.

———. 1994. *Anthropology of the Self*. London: Pluto Press.

Murdoch, Iris. 1992. *Metaphysics as a Guide to Morals*. Harmondsworth: Penguin.

Nencel, Lorraine, and Peter Pels. 1991. *Constructing Knowledge*. London: Sage.

O'Meara, J. Tim. 1989. "Anthropology as Empirical Science." *American Anthropologist* 91: 354–69.

O'Neill, John. 1995. *The Poverty of Postmodernism*. London: Routledge.

Outhwaite, William. 1975. *Understanding Social Life: The Method Called Verstehen*. London: Allen & Unwin.

———. 1987. *New Philosophies of Social Science*. London: MacMillan.

———. 1994. *Habermas: A Critical Introduction*. Cambridge: Polity Press.

Passmore, John Arthur. 1985. *Recent Philosophers*. London: Duckworth.

Pennick, Nigel. 1996. *Celtic Sacred Landscapes*. London: Thames & Hudson.

Pocock, David, F. 1961. *Social Anthropology*. London: Steed & Ward.

Popper, Karl. 1992a. *In Search of a Better World*. London: Routledge.

———. 1992b. *Unended Quest*. London: Routledge.

Rabinow, Paul. 1984. *The Foucault Reader*. Harmondsworth: Penguin.

Rabinow, Paul, and William M. Sullivan, eds. 1987. *Interpretive Social Science: A Second Look*. Berkeley: University of California Press.

Radcliffe-Brown, A.R. 1922. *The Andaman Islanders*. Glencoe: Free Press.

Reyna, S.P. 1994. "Literary Anthropology and the Case against Science." *Man* 29, no. 3: 555–81.

Ricoeur, Paul. 1976. *Interpretation Theory: Discourse and the Surplus of Meaning*. Fort Worth: Texas Christian University Press.

———. 1981. *Hermeneutics and the Human Sciences*. Cambridge University Press.

Rorty, Richard. 1979. *Philosophy and the Mirror of Nature*. Princeton, NJ: Princeton University Press.

———. 1989. *Contingency, Irony and Solidarity*. Cambridge University Press.

Rosenau, Pauline M. 1992. *Postmodernism and the Social Sciences*. Princeton, NJ: Princeton University Press.

Shweder, Richard A. 1984. "Anthropology's Romantic Rebellion against the Enlightenment." In *Culture Theory: Essays on Mind, Self, and Emotion*, edited by Richard A. Shweder and Robert A. LeVine, 27–66. Cambridge University Press.

Skinner, B.F. 1953. *Science and Human Behaviour*. New York: MacMillan.

Solomon, Robert C. 1988. *Continental Philosophy since 1750*. New York: Oxford University Press.

Sorell, Tom. 1991. *Scientism: Philosophy and the Infatuation with Science*. London: Routledge.

Sperber, Dan. 1985. *On Anthropological Knowledge*. Cambridge: Cambridge University Press.

Taylor, Charles. 1985. *Philosophy and the Human Sciences 2*. Cambridge: Cambridge University Press.

Toulmin, Stephen. 1995. Foreword. In *Rethinking Knowledge: Reflections across the Disciplines*, edited by Robert F. Goodman and Walter R. Fisher. *Rethinking Knowledge*. Albany: State University of New York Press.

Trigg, Roger. 1980. *Reality at Risk*. Hemel Hempstead: Harvester.

Tyler, Stephen. 1973. *India: An Anthropological Perspective*. Pacific Palisades, CA: Goodyear Publishing.

———. 1986. "Postmodern Ethnography." In *Writing Culture: The Poetics and Politics of Ethnography*, edited by James Clifford and George Marcus, 122–40. Berkeley: University of California Press.

———. 1991. "A Postmodern Instance." In *Constructing Knowledge*, edited by Lorraine Nencel and Peter Pels, 78–94. London: Sage.

Vincent, Joan. 1990. *Anthropology and Politics: Visions, Traditions, and Trends*. Tucson: University of Arizona Press.

Warnke, Georgia. 1987. *Gadamer: Hermeneutics, Tradition, and Reason*. Cambridge: Polity Press.

Watkins, J.W.N. 1968. "Methodological Individualism and Social Tendencies." In *Readings in the Philosophy of the Social Sciences*, edited by May Brodbeck. New York: MacMillan.

White, Leslie A. 1949. *The Science of Culture*. New York: Grove Press.

Wolin, Richard. 1990. *The Politics of Being: The Political Thought of Martin Heidegger*. New York: Columbia University Press.

3

Anthropology and Anarchism (1998)

There is, in many ways, an "elective affinity" between anthropology and anarchism. Although anthropology's subject matter has been diverse, and its conspectus rather broad—as a study of human culture—historically it has always had a rather specific focus—on the study of pre-state societies. But it is quite misleading to portray the anthropology of the past as being simply the study of so-called "primitive" people or the "exotic" other, and thus largely engaged in a kind of "savage" operation of "disappearing" cultures. This is a rather biased and inaccurate portrait of anthropology, for the discipline has a long tradition of "anthropology at home," and many important anthropological studies have their location in India, China, and Japan. It is thus noteworthy that James Clifford and George Marcus (1986) in what many have regarded as the founding text of literary or postmodern anthropology, are not only rather dismissive of feminist anthropology, but ignore entirely the ethnographic studies of non-"Western" scholars—Srinivas, Kenyatta, Fei, and Aiyappan. But in an important sense anthropology is the social science discipline that has put a focal emphasis on those kinds of societies that have been seen as exemplars of anarchy, a society without a state. Indeed, Evans-Pritchard, in his classic study of the Nuer (1940), described their political system as "ordered anarchy." Harold Barclay's useful and perceptive little book *People without government* (1992) is significantly subtitled *The Anthropology of Anarchism*, and Barclay makes the familiar distinction between anarchy, which is an ordered society without government, and anarchism, which is a political movement and tradition that became articulated during the nineteenth century.

Anthropologists & Anarchism: Reclus, Bouglé, Mauss, Radcliffe-Brown

What I want to do in this article is to explore some connections between anarchism and anthropology, by looking at some anthropologists who had relations with the anarchist tradition, by briefly examining how anarchists have drawn upon anthropological writings, and will conclude by offering some personal reflections on anarchism.

Many anthropologists have had affinities with anarchism. One of the earliest ethnographic texts was a book by Élie Reclus called *Primitive Folk*. It was published in 1903, and carries the subtitle *Studies in Corporative Ethnology*. It is based on information derived from the writings of travellers and missionaries, and it has the evolutionary flavour of books written at the end of the nineteenth century, but it contains lucid and sympathetic accounts of such people as the Apaches, Nayars, Todas, and Inuits. Reclus declares the moral and intellectual equality of these cultures with that of "so-called civilised states," and it is of interest that Reclus used the now familiar term *Inuit*, which means "people," rather than the French term Eskimo. Élie Reclus was the elder brother, and lifetime associate, of Elisée, the more famous anarchist-geographer.

Another French anthropologist with anarchist sympathies was Célestin Bouglé, who wrote not only a classical study of the Indian caste system (1908)—which had a profound influence on Louis Dumont—but also an important study of Proudhon. Bouglé was one of the first to affirm, then (1911) controversially, that Proudhon was a sociological thinker of standing. There was in fact a close relationship between the French sociological tradition, focused around Durkheim, and both socialism and anarchism, even though Durkheim himself was antagonistic to the anarchist stress on the individual. Durkheim was a kind of guild socialist, but his nephew Marcel Mauss wrote a classical study, *The Gift* (1925), which focused on reciprocal or gift exchange among pre-literate cultures. This small text is not only in some ways an anarchist tract, it is one of the foundation texts of anthropology, one read by every budding anthropologist. British anthropologists have less connection with anarchism, but it is worth noting that one of the so-called "fathers" of British anthropology, A.R. Radcliffe-Brown was an anarchist in his early years.

Alfred Brown was a lad from Birmingham. He managed, with the help of his brother, to get to Oxford University. There two influences were important to him. One was the process philosopher Alfred Whitehead, whose organismic theory had a deep influence on Radcliffe-Brown. The other was Kropotkin, whose writings he imbibed. In his student days at Oxford Radcliffe-Brown was known as "Anarchy Brown." Alas! Oxford got to him. He later became something of an intellectual aristocrat and changed his name to the hyphenated "A.R. Radcliffe-Brown." But, as Tim

Ingold has written (1986), Radcliffe-Brown's writings are permeated with a sense that social life is a process, although like most Durkheimian functionalists he tended to play down issues relating to conflict, power, and history.

Although anarchism has had a minimal influence on anthropology—though many influential anthropologists can be described as radical liberals and socialists (like Boas, Radin, and Diamond)—anarchist writers have drawn extensively on the work of anthropologists. Indeed there is a real contrast between anarchists and Marxists with respect to anthropology, for while anarchists have critically engaged themselves with ethnographic studies, Marxist attitudes to anthropology have usually been dismissive. In this respect Marxists have abandoned the broad historical and ethnographic interests of Marx and Engels. The famous study of Engels's, *The Origin of the Family, Private Property and the State* (1884), is of course based almost entirely on Lewis Morgan's anthropological study of *Ancient Society* (1877). If one examines the writings of all the classical Marxists—Lenin, Trotsky, Gramsci, Lukacs—they are distinguished by a wholly Eurocentric perspective, and a complete disregard for anthropology. The entry under "Anthropology" in *A Dictionary of Marxist Thought* (Bottomore 1983), significantly has nothing to report between Marx and Engels in the nineteenth century, and the arrival on the scene of French Marxist anthropologists in the 1970s (Godelier, Meillassoux). Equally amazing is that one Marxist text, specifically *Pre-Capitalist Modes of Production* (Hindness and Hirst 1975), not only suggested that the "objects" of theoretical discourses did not exist—and so rejected history as a worthwhile subject of study—but completely bypassed anthropological knowledge. This is matched of course by the dismissive attitude towards anarchism by Marxist scholars—Perry Anderson, Wallerstein, and E.P. Thompson are examples.

Anarchists & Anthropology: Kropotkin, Bookchin, Clastres, Zerzan

In examining those anarchists who have creatively utilised anthropology, I will discuss briefly only four writers—Kropotkin, Bookchin, Clastres, and Zerzan.

Kropotkin is well known. But being a geographer as well as an anarchist, and having travelled widely in Asia, Kropotkin had wide ethnographic interests. This is most clearly expressed in his classic text *Mutual Aid* published in 1903. In this book Kropotkin attempted to show that organic and social life were not arenas where laissez-faire competition and conflict and the "survival of the fittest" were the only norms, but rather these domains were characterised by "mutuality" and "symbiosis." It was the ecological dimension of Darwin's thought, expressed in the last chapter of *The Origin of Species*, that was crucial for Kropotkin; cooperation, not struggle, was the important factor in the evolutionary process. This is exemplified by the ubiquitous lichen, one of the most basic forms of life and found practically everywhere.

Kropotkin's book gives lengthy accounts of mutual aid not only among hunter-gatherers and such people as the Buryat and Kabyle (now well-known through Bourdieu's writings), but also in the medieval city and in contemporary European societies. In an ASA monograph on socialism (Hann 1993), two articles specifically examine anarchy among contemporary people. Alan Barnard looks at the issues of "primitive communism" and "mutual aid" among the Kalahari hunter-gatherers, while Joanna Overing discusses "anarchy and collectivism" among the horticultural Piaroa of Venezuela. Barnard's essay has the subtitle "Kropotkin visits the Bushmen," indicating that anarchism is still a live issue among some anthropologists.

Kropotkin was concerned to examine the "creative genius" of people living at what he described as the "clan period" of human history, and the development of institutions of mutual aid. But this did not entail the repudiation of individual self-assertion, and, unlike many contemporary anthropologists, Kropotkin made a distinction between individuality and self-affirmation, and individualism.

Murray Bookchin is a controversial figure. His advocacy of citizens' councils and municipal self management, his emphasis on the city as a potential ecological community, and his strident critiques of the misanthropy and eco-mysticism of the deep ecologists are perhaps well known, and the centre of many debates—much of it acrimonious. But Bookchin's process-oriented dialectical approach, and his sense of history—alive to the achievements of the human spirit—inevitably led Bookchin to draw on anthropological studies. The main influences on his work were Paul Radin and Dorothy Lee, both sensitive scholars of native American culture. In *The Ecology of Freedom* (1982), Bookchin devotes a chapter to what he describes as "organic society," emphasising the important features of early human tribal society, a primordial equality and the absence of coercive and domineering values, a feeling of unity between the individual and the kin community, a sense of communal property and an emphasis on mutual aid and usufruct rights, and a relationship with the natural world which is one of reciprocal harmony, rather than of domination. But Bookchin is concerned that we draw lessons from the past, and learn from the culture of pre-literate people, rather than romanticising the life of hunter-gatherers. Still less, that we should try to emulate them.

Pierre Clastres was both an anarchist and an anthropologist. His minor classic, on the Indian communities of South America—specifically the forest Guayaki (Aché)—is significantly titled "Society *against* the State"

(1977). Like Tom Paine and the early anarchists, Clastres makes a clear distinction between society, as a pattern of social relations, and the state, and argues that the essence of what he describes as "archaic" societies— whether hunter-gatherers or horticultural (neolithic) peoples—is that effective means are institutionalised to prevent power being separated from social life. He bewails the fact that Western political philosophy is unable to see power except in terms of "hierarchized and authoritarian relations of command and obedience" (9) and thus equates power with coercive power. Reviewing the ethnographic literature of the people of South America—apart from the Inca State—Clastres argues that they were distinguished by their "sense of democracy and taste for equality," and that even local chiefs lacked coercive power. What constituted the basic fabric of archaic society, according to Clastres, was that of exchange, coercive power, in essence, being a negation of reciprocity. He contends that the aggressiveness of tribal communities has been grossly exaggerated, and that a subsistence economy did not imply an endless struggle against starvation, for in normal circumstances there was an abundance and variety of things to eat. Such communities were essentially egalitarian, and people had a high degree of control over their own lives and work activities. But the decisive "break" for Clastres, between "archaic" and "historical" societies, was not the neolithic revolution and the advent of agriculture, but the "political revolution" involving the intensification of agriculture and the emergence of the state.

The key points of Clastres's analysis have recently been affirmed by John Gledhill (1994, 13–15); it provides a valuable critique of Western political theory which identifies power with coercive authority, and it suggests looking at history less in terms of typologies but as a historical process in which human activities have endeavoured to maintain their own autonomy, and to resist the centralising intrusions and the exploitation inherent in the state.

While for Clastres and Bookchin political domination and hierarchy begin with the intensification of agriculture, and the rise of the state, for John Zerzan the domestication of plants and animals heralds the demise of an era when humans lived an authentic free life. Agriculture, per se, is a form of alienation; it implies a loss of contact with the world of nature and a controlling mentality. The advent of agriculture thus entails the "end of innocence" and the demise of the "golden age" as humans left the "garden of eden," though Eden is identified not with a garden but with hunter-gathering existence. Given this advocacy of "primitivism" it is hardly surprising that Zerzan (1988, 1994) draws on anthropological data to validate his claims, and to portray hunter-gatherers as egalitarian, authentic, and as the "most successful and enduring adaptation ever achieved by

humankind" (1988, 66). Even symbolic culture and the shamanism associated with hunter-gatherers is seen by Zerzan as implying an orientation
to manipulate and control nature or other humans. Zerzan presents an
apocalyptic, even a gnostic vision; our hunter-gatherer past is described
as an idyllic era of virtue and authentic living; the last eight thousand
years or so of human history—after the fall (agriculture)—is seen as one
of tyranny, hierarchic control, mechanised routine devoid of any spontaneity, and as involving the anesthetisation of the senses. All those products
of the human creative imagination—farming, art, philosophy, technology, science, urban living, symbolic culture, are all viewed negatively by
Zerzan—in a monolithic sense. The future we are told is "primitive"; how
this is to be achieved in a world that currently sustains almost six billion
people, for evidence suggests that the hunter-gather lifestyle is only able to
support one or two people per square mile; or whether the "future primitive" actually entails, in gnostic fashion, a return not to the godhead, but to
hunter-gathering subsistence, Zerzan does not tell us. While radical ecologists glorify the golden age of peasant agriculture, Zerzan follows the likes
of Van Der Post in extolling hunter-gatherer existence—with a selective
culling of the anthropological literature. Whether such "illusory images of
Green primitivism" are, in themselves, symptomatic of the estrangement
of affluent urban dwellers and intellectuals from the natural (and human)
world—as both Bookchin (1995) and Ray Ellen (1986) suggest—I will leave
to others to judge.

Reflections on Anarchism

The term anarchy comes from the Greek, and essentially means "no
ruler." Anarchists are people who reject all forms of government or coercive authority, all forms of hierarchy and domination. They are therefore
opposed to what the Mexican anarchist Flores Magón called the "sombre
trinity"—state, capital, and the church. Anarchists are thus opposed to
both capitalism and to the state, as well as to all forms of religious authority. But anarchists also seek to establish or bring about by varying means a
condition of anarchy, that is, a decentralised society without coercive institutions, a society organised through a federation of voluntary associations.
Contemporary right-wing libertarians, like Milton Friedman, Rothbard,
and Ayn Rand, who are often described as "anarcho-capitalists," and who
fervently defend capitalism, are not in any real sense anarchists.

In an important sense anarchists support the rallying cry of the French
Revolution—liberty, equality, and fraternity—and strongly believe that
these values are interdependent. As Bakunin remarked: "Freedom without
socialism is privilege and injustice; and socialism without freedom is
slavery and brutality." Needless to say anarchists have always been critical

of Soviet communism, and the most powerful and penetrating critiques of Marx, Marxist-Leninism, and the Soviet regime have come from anarchists: people like Berkman, Goldman, and Maximoff. The latter's work was significantly titled *The Guillotine at Work* (1940). Maximoff saw the politics of Lenin and Trotsky as similar to that of the Jacobins in the French Revolution, and equally reactionary.

With the collapse of the Soviet regime, Marxists are now in a state of intellectual disarray, and are floundering around looking for a safe political anchorage. They seem to gravitate either towards Hayek or towards Keynes; whichever way, their socialism gets lost in the process. Conservative writers like Roger Scruton take great pleasure in berating Marxists for having closed their eyes to the realities of the Soviet regime; they themselves, however, have a myopia when it comes to capitalism. The poverty, famine, sickening social inequalities, political repression, and ecological degradation that is generated under capitalism is always underplayed by apologists like Scruton and Fukuyama. They see these as simply "problems" that need to be overcome—not as intrinsically related to capitalism itself.

Anarchism can be looked at in two ways.

On the one hand it can be seen as a kind of "river," as Peter Marshall describes it in his excellent history of anarchism (2010). It can thus be seen as a "libertarian impulse" or as an "anarchist sensibility" that has existed throughout human history: an impulse that has expressed itself in various ways—in the writings of Lao Tzu and the Taoists, in classical Greek thought, in the mutuality of kin-based societies, in the ethos of various religious sects, in such agrarian movements as the Diggers in England and the Zapatistas of Mexico, in the collectives that sprang up during the Spanish Civil War, and, currently, in the ideas expressed in the ecology and feminist movements. Anarchist tendencies seem to have expressed themselves in all religious movements, even in Islam. One Islamic sect, the Najadat, believed that "power belongs only to god." They therefore felt that they did not really need an imam or caliph, but could organise themselves mutually to ensure justice. Many years ago I wrote an article on Lao Tzu, suggesting that the famous *Tao Te Ching* ("The Way and Its Power," as Waley translates it) should not be seen as a mystical religious tract (as it is normally understood), but rather as a political treatise. It is, in fact, the first anarchist tract. For the underlying philosophy of *Tao Te Ching* is fundamentally anarchist, as Rudolf Rocker long ago noted.

On the other hand anarchism may be seen as a historical movement and political theory that had its beginnings at the end of the eighteenth century. It was expressed in the writings of William Godwin, who wrote the classic anarchist text *An Enquiry Concerning Political Justice* (1793), as well as in the actions of the *sans-culottes* and the *enragés* during the French Revolution and by radicals like Thomas Spence and William Blake in Britain. The term "anarchist" was first used during the French Revolution as a term of abuse in describing the sans-culottes—"without breeches"— the working people of France who during the revolution advocated the abolition of government.

Anarchism, as a social movement, developed during the nineteenth century. Its basic social philosophy was formulated by the Russian revolutionary Michael Bakunin. It was the outcome of his clashes with Karl Marx and his followers—who advocated a statist road to socialism—during meetings of the International Working Men's Association in the 1860s. In its classical form, therefore, as it was expressed by Kropotkin, Goldman, Reclus, and Malatesta, anarchism was a significant part of the socialist movement in the years before the First World War, but its socialism was libertarian not Marxist. The tendency of writers like David Pepper (1996) to create a dichotomy between socialism and anarchism is, I think, both conceptually and historically misleading.

Of all political philosophies anarchism has had perhaps the worst press. It has been ignored, maligned, ridiculed, abused, misunderstood, and misrepresented by writers from all sides of the political spectrum— Marxists, liberals, democrats, and conservatives. Theodore Roosevelt, the American president, described anarchism as a "crime against the whole human race"—and it has been variously judged as destructive, violent, and nihilistic. A number of criticisms have been lodged against anarchism, and I will deal briefly with each of these. There are eight complaints in all.

1. It is said that anarchists are too innocent, too naive, and have too rosy a picture of human nature. It is said that, like Rousseau, they have a romantic view of human nature which they see as essentially good and peace-loving. But of course real humans are not like this; they are cruel and aggressive and selfish, and so anarchy is just a pipe dream. It is an unrealistic vision of a past golden age that never really existed. This being so, some form of coercive authority is always necessary. The truth is that anarchists do not follow Rousseau. In fact, Bakunin was scathing in his criticisms of the eighteenth-century philosopher. Most anarchists tend to think humans have both good and bad tendencies. If they did think humans all goodness and light, would they mind being ruled? It is because they have a realistic rather than a romantic view of human nature, that they oppose all forms of coercive authority. In essence, anarchists oppose all power which the French describe as *puissance*—"power over" (rather than *pouvoir*, the power to do something) and believe, like Lord Acton, that power corrupts, and absolute power corrupts absolutely. As Paul Goodman wrote: "the issue is not whether people are 'good enough' for a particular

type of society; rather it is a matter of developing the kind of social institutions that are most conducive to expanding the potentialities we have for intelligence, grace, sociability and freedom."

2. Anarchy, it is believed, is a synonym for chaos and disorder. This is, in fact, how people often use the term. But anarchy, as understood by most anarchists, means the exact opposite of this. It means a society based on order. Anarchy means not chaos, or a lack of organisation, but a society based on the autonomy of the individual, on cooperation, one without rulers or coercive authority. As Proudhon put it: liberty is the mother of order. But equally anarchists do not denounce chaos, for they see chaos and disorder as having inherent potentiality; as Bakunin put it: to destroy is a creative act.

3. Another equation made is that between anarchism and violence. Anarchism, it is said, is all about terrorist bombs and violence. And there is a book currently in the bookshops entitled *The Anarchist Cookbook*. It is all about how to make bombs and dynamite. But as Alexander Berkman wrote: the resort to violence against oppression or to obtain certain political objectives has been practiced throughout human history. Acts of violence have been committed by the followers of every political and religious creed: nationalists, liberals, socialists, feminists, republicans, monarchists, Buddhists, Muslims, Christians, democrats, conservatives, fascists . . . and every government is based on organised violence. Anarchists who have resorted to violence are no worse than anybody else. But most anarchists have been against violence and terrorism, and there has always been a strong link between anarchism and pacifism. Yet anarchists go one step further: they challenge the violence that most people do not recognise and which is often of the worst possible kind; this is lawful violence. Needless to say, some of the best-known anarchists, like Tolstoy, De Cleyre, Gandhi, and Edward Carpenter, were pacifists.

4. Anarchists have been accused, especially by Marxists, of being theoretical blockheads, of being anti-intellectual, or of making a cult of mindless action. But as a perusal of the anarchist movement will indicate, many anarchists or people with anarchist sympathies have been among the finest intellects of their generation, truly creative people. One may mention: Godwin, Humboldt, Reclus, Tolstoy, Russell, Gandhi, Chomsky, Bookchin. Moreover, anarchists have produced many seminal texts outlining their own philosophy and their own social doctrines. These are generally free of the jargon and the pretension that masks as scholarship among many liberal scholars, Marxists, and postmodernists.

5. Another criticism is the opposite of this: it ridicules anarchism for being apolitical and a doctrine of inaction. Anarchists, according to the ex-doyen of the Green Party Jonathon Porritt, do nothing but contemplate

their navels. Because they do not engage in party politics, he even suggests that anarchists do not live in the "real world." All the essential themes of the Green Party manifesto—the call for a society that is decentralised, equitable, ecological, cooperative, with flexible institutions—are of course simply an unacknowledged appropriation of what anarchists like Kropotkin had long ago advocated—but with Porritt this vision is simply hitched to party politics. As a media figure Porritt completely misunderstands what anarchism—and a decentralised society—is all about. Anarchism is not nonpolitical. Nor does it advocate a retreat into prayer, self-indulgence, or meditation, whether or not one contemplates one's navel or chants mantras. It is simply hostile to parliamentary or party politics. The only democracy it thinks valid is participatory democracy, and it considers putting an X on a piece of paper every five years a sham. It serves only to give ideological justification to power holders in a society that is fundamentally hierarchic and undemocratic. Anarchists are of many kinds. They have therefore suggested various ways of challenging and transforming the present system of violence and inequality—through communes, passive resistance, syndicalism, municipal democracy, insurrection, direct action, and education. One of the reasons why some anarchists have put a lot of emphasis on publishing propaganda and education is that they have always eschewed party organisation as well as violence. Anarchists have always been critical of the notion of a vanguard party, seeing it as inevitably leading to some form of despotism. And with regard to both the French and Russian revolutions, history has proved their premonitions correct.

6. A consistent critique of anarchism offered by Marxists is that it is utopian and romantic, a peasant or petit-bourgeois ideology, or an expression of millennial dreams. Concrete historical studies by John Hart on anarchism and the Mexican working class (1978) and by Jerome Mintz on the anarchists of Casas Viejas in Spain (1982) have more than adequately refuted some of the distortions about anarchism. The anarchist movement has not been confined to peasants: it has flourished among urban workers where anarcho-syndicalism developed. Nor is it utopian or millennial. Anarchists have established real collectives and have always been critical of religion. Nobody among the early anarchists expected some immediate or cataclysmic change to occur through "propaganda by deed" or the "general strike"—as the writings of Reclus and Berkman attest. They realised it would be a long haul.

7. Another criticism of anarchism is that it has a narrow view of politics: that it sees the state as the fount of all evil, ignoring other aspects of social and economic life. This is a misrepresentation of anarchism. It partly derives from the way anarchism has been defined, and partly because Marxist historians have tried to exclude anarchism from the

broader socialist movement. But when one examines the writings of classical anarchists like Kropotkin, Goldman, Malatesta, and Tolstoy, as well as the character of anarchist movements in such places as Italy, Mexico, Spain, and France, it is clearly evident that it has never had this limited vision. It has always challenged all forms of authority and exploitation, and has been equally critical of capitalism and religion as it has of the state. Most anarchists were feminists, and many spoke out against racism, as well as defending the freedom of children. A cultural and ecological critique of capitalism has always been an important dimension of anarchist writings. This is why the writings of Tolstoy, Reclus, and Kropotkin have contemporary relevance.

8. A final criticism of anarchism is that it is unrealistic: anarchy will never work. The market socialist David Miller expresses this view very well in his book *Anarchism* (1984). His attitude to anarchism is one of heads I win, tails you lose. He admits that communities based on anarcho-communist principles have existed, and "given a chance" have had some degree of "unexpected success." But due to lack of popular support and state intervention and repression they have, he writes, always been "failures." On the other hand he also argues that societies could not exist anyway without some form of centralised government. Miller seems oblivious to the fact that what Stanley Diamond called "kin-communities" have long existed within and often in opposition to state systems, and that trading networks have existed throughout history, even among hunter-gatherers, without any state control. The state, in any case, is a recent historical phenomena, and in its modern nation-state form has only existed for a few hundred years. Human communities have long existed without central or coercive authority. Whether a complex technological society is possible without centralised authority is not a question easily answered; neither is it one that can be lightly dismissed. Many anarchists believe that such a society is possible, though technology will have to be on a "human scale." Complex systems exist in nature without there being any controlling mechanism. Indeed, many global theorists nowadays are beginning to contemplate libertarian social vistas that become possible in an age of computer technology. Needless to say, if Miller had applied the same criteria by which he so adversely adjudges anarchism—distributive justice and social well-being—to capitalism and state "communism" then perhaps he would have declared both these systems unpractical and unrealistic too. But at least Miller wants to rescue anarchism from the dustbin of history—to help us to curb abuses of power, and to keep alive the possibilities of free social relationships.

Society, we are told by such authorities as Friedrich Hayek, Margaret Thatcher, and Marilyn Strathern, either does not exist or it is a "confused category" that ought to be excised from theoretical discourse. The word derives, of course, from the Latin, *Societas*, which in turn derives from *Socius*, meaning a companion, a friend, a relationship between people, a shared activity. Anarchists have thus always drawn a clear distinction between society, in this sense, and the state: between what the Jewish existentialist scholar Martin Buber called the "political" and the "social" principles. Buber was a close friend of the anarchist Gustav Landauer, and what Landauer basically argued—long before Foucault—was that the state could not be destroyed by revolution: it could only be undermined—by developing other kinds of relationships, by actualising social patterns and forms of organisation that involved mutuality and free cooperation. Such a social domain is always in a sense present, imminent in contemporary society, coexisting with the state. For Landauer, as for Colin Ward, anarchy, therefore, is not something that only existed long ago before the rise of the state, or exists now only among people like the Nharo or Piaroa living at the margins of capitalism (although Zerzan would not see the Piaroa, being horticulturists, as truly authentic!), nor is it simply a speculative vision of some future society; rather, anarchy is a form of social life which organises itself without the resort to coercive authority. It is always in existence—albeit often buried and unrecognised beneath the weight of capitalism and the state. It is like "a seed beneath the snow" as Colin Ward (1973) graphically puts it. Anarchy, then, is simply the idea, to stay with the same writer, "that it is possible and desirable for society to organise itself without government."

References

Barclay, Harold. 1982. *People without Government*. London: Kahn & Averill.

Bookchin, Murray. 1982. *The Ecology of Freedom*. Palo Alto: Cheshire Books.

———. 1995. *Re-enchanting Humanity*. London: Cassell.

Bottomore, Tom, ed. 1983. *A Dictionary of Marxist Thought*. Oxford: Blackwell.

Clastres, Pierre. 1977. *Society against the State*. Oxford: Blackwell.

Clifford, James, and George E. Marcus, eds. 1986. *Writing Culture: The Poetics and Politics of Ethnography*. Berkeley: University of California Press.

Ellen, Roy F. 1986. "What Black Elk Left Unsaid." *Anthropology Today* 216: 8–12.

Gledhill, John. 1994. *Power and Its Disguises: Anthropological Perspectives on Politics*. Condon: Pluto Press.

Hann, Chris M. 1993. *Socialism: Ideals, Ideologies and Local Practice*. London: Routledge.

Hart, John M. 1978. *Anarchism and the Mexican Working Class, 1860–1931*. Austin: University of Texas.

Hindness, Barry, and Paul Q. Hirst. 1975. *Pre-Capitalist Modes of Production*. London: Routledge & Kegan Paul.

Ingold, Tim. 1986. *Evolution and Social Life*. Cambridge: Cambridge University Press.

Marshall, Peter. 2010. *Demanding the Impossible: A History of Anarchism*. Oakland: PM Press.

Miller, David. 1984. *Anarchism*. London: Dent.

Mintz, Jerome R. 1982. *The Anarchists of Casas Viejas*. Chicago: University of Chicago Press.
Pepper, David. 1996. *Modern Environmentalism*. London: Routledge.
Ward, Colin. 1973. *Anarchy in Action*. London: Allen & Unwin.
Zerzan, John. 1988. *Elements of Refusal*. Seattle: Left Bank Books.
_____. 1994. *Future Primitive and Other Essays*. Brooklyn: Autonomedia.

4

Buddhism, Anarchy, and Ecology (1999)

Prologue

This article was rejected by the journal *Anarchist Studies* because although considered thought-provoking it was said to be a one-sided polemic that did not fully explore the relationship between Buddhism and anarchism (and ecology) as suggested by the likes of Watts, Snyder, and Marshall. In fact the aim was not to explore the relationship between Buddhism and anarchism but—as explicitly stated—to offer critical reflections on Buddhism from an anarchist perspective and thus counter the biased, misleading, and one-sided presentation of Buddhism by these writers and by eco-Buddhists, and to draw attention to the more negative aspects of Buddhism, particularly its symbiotic relationship with state structures.

In recent decades Buddhism has become increasingly popular among people in both Europe and the United States, and societies such as the "Friends of the Western Buddhist Order" now have centres in all the major cities. On the surface, of all the major world religions Buddhism would seem to be the closest to the anarchist tradition, and in recent years there have been a spate of articles and books claiming that Buddhism entails an environmental ethic and is an eco-philosophy par excellence.[1] In this paper I offer some critical reflections on Buddhism from an anarchist perspective. The paper is in two parts. In the first part I offer a brief overview of Buddhism as a religious philosophy, stressing that in essence Buddhism is a way of salvation, a radical form of mysticism that demands detachment from the empirical world. In the second part I note briefly some of the limitations of Buddhism, its symbiotic relationship with state power and the anti-ecological tenets of canonical Buddhism as a practice.[2]

1. The Dharma as Understood by Buddhist Scholars

It is important to recognise that the Buddha—the historical Buddha, Siddhartha Gautama, who lived in the sixth century BC and who is seen as the founder of Buddhism—at no time claimed to be anything other than a human being. He claimed no divine inspiration, nor any revelation. What he did claim was to have become awakened or enlightened. Buddha means one who has become enlightened—the "awakened one."[3]

Buddhists are essentially those who follow the path of the historical Buddha, and his teachings or truths are known as the dharma. The dharma, the "way of enlightenment" was compared by the Buddha to a raft, which should be discarded when the river was crossed.[4]

The dharma, as Sangharakshita (Dennis Lingwood) has insisted,[5] consists of anything that is conducive to a person's enlightenment. As the Buddha said: "Whosoever teachings conduce to dispassion, to detachment, to decrease in worldly gains, to frugality, to content, to solitude, to energy, to delight in good, of these teachings you can be certain that they are the teaching of the Buddha" (Sangharakshita 1980, 27).

Buddhism is therefore a practical religion and in its canonical form is a religion of tolerance and emphasises the basic goodness of humans, unlike orthodox Christianity. There is no emphasis on faith and in no sense is the Buddha, or the scriptures, to be taken as an incontrovertible source of truth. On his death bed the Buddha is reported to have told his disciple Ananda that decay is inherent in all conditioned things, that he should not fret or weep at his passing and that his disciples should seek their own salvation. "Be a refuge unto yourselves," he said.[6] In this sense Buddha was an anarchist.

Buddhism is non-theistic, indeed it has often been described as atheism, for the belief in a divine creator or in a personal deity is seen as essentially a hindrance to the spiritual life (salvation). The Buddha also expressed a disinterest in philosophical speculation or in obtaining occult powers. According to Walpola Rahula's interpretation of canonical Buddhism the Buddha completely undermined two ideas that are deeply rooted in the human psyche—the need for self-protection and the need for self-preservation. For self-protection humans in their imagination create the idea of religion, of a god or spirits also will protect them and keep them well and secure. For self-preservation they conceive the idea of an immortal soul that survives death. Yet according to the Buddha these two ideas are false, empty, and give illusionary comfort. They are simply born out of ignorance and an unwillingness on the part of humans to face reality.[7]

The aim of life for a Buddhist is a spiritual one, to achieve enlightenment (nirvana).[8] This is seen, in its simplest form, as the "extinction" of desire (passion, craving, becoming), of aversion (hatred, anger), and of confusion (delusion, ignorance). In the Buddhist scriptures and iconography the three "mental poisons" to be extinguished—desire, aversion, ignorance—are represented by three animals respectively, the cock (a sexual symbol), the snake, and the pig (wallowing ignorantly in the mud).

Buddhist cosmology, derived from early Hinduism suggests that the cosmos forms a "wheel" of life, consisting of six realms of existence—the gods (devas), the spirits (asuras), ghosts of the dead (pretax) [relating to hot and cold realms], humans, and animals. Together they constitute a cycle of rebirths (samsara). The "wheel of life," in more metaphysical terms, can also be seen as a chain of "dependent origination" (paticca samuppada), the notion that everything arises from causes and conditions, and, thus, that everything in the world is conditioned, relative, interdependent, and in a state of flux.

Not being a revealed religion, not being, that is, a religion that is based on revelation or faith, in God or in some divine message—as with Judaism, Christianity, Islam, and to some extent Hinduism—Buddhism puts a fundamental emphasis on human experience. It has often been described as a form of radical empiricism (Conze 1951, 20). It has thus, like Taoism, been characterised as a "religion-of-discovery," a manifestation of the human spirit, truth being discovered by humans, by their own unaided effort (Sangharakshita 1987, 148–50). The Buddha therefore is not conceived of as a prophet, still less as God incarnate, but simply as a teacher, guide, or exemplar. He described himself, on this regard, as like the first chick to hatch from a batch of eggs (Conze 1954, 60; Sangharakshita 1980, 25).

The four noble truths (arya satya) that the Buddha expounded in the deer park near Benares (Varanasi) (in around 531 BC), constitute the essence of the dharma and are as follows:

1. Everything is suffering (dukkha);
2. The origin of suffering is desire (tanha);
3. There exists an end to suffering; and
4. A path is suggested by the Buddha that leads to salvation (nirvana).

The key idea in this is that the world is bound up with suffering. It is not a matter of original sin—life, existence is suffering. As the Buddha said in one of his discourses: "Birth is suffering; old age is suffering; death is suffering; sorrow, lamentation, pain, grief, and despair are suffering; not to get what are wishes is suffering; in short the five aggregates of clinging are suffering" (Nyanaponika Thera 1962, 127).

The Buddha, of course, did not deny the enjoyment and happiness, both material and spiritual, that there is in life, but stressed that these were transient and impermanent. Nor did this emphasis on the reality of suffering suggest a rejection of the world—a repugnant attitude towards the material world. For the Buddha, suffering was an existential fact, not an evil, and there is nothing to be gained by expressing anger, gloom, or impatience about it. According to the Buddha one of the principal hindrances to salvation is repugnance or hatred. Thus what is necessary, according to the Buddha, is not anger or pessimism, but the understanding of suffering, how it comes about, and how to get rid of it. Rahula notes how significant it is that the Buddha is always depicted as a happy person, "ever-smiling," always serene, contented, tranquil, and compassionate (1959, 27–28).

Most Buddhist scholars have suggested or implied that suffering is an intrinsic property of the world, that it thus has an ontological status. But David Shaner writes: "'Suffering' in the Buddhist sense is not a metaphysical or ontological declaration concerning a pessimistic human condition; it is rather an epistemologically relevant term that defines the perspective of those whose perception of nature is clouded by desires, false self-images, and the like" (1989, 170).

"Suffering" is thus interpreted not as an existential fact, but as a mental disposition, an intentional condition due to excessive desire and covetousness.[9] If suffering thus arises out of our desires and our craving for stability and permanence then, as the Buddha seems to suggest, extinguishing such desires will lead to salvation as he conceived it. This we discuss below. The concept of dukkha—a term which not only means pain and misery but also imperfection, disharmony, and discontent—is intrinsically connected in Buddhist thought to two other ideas, or "characteristics" (laksanas) of existence, that of impermanence (anicca) and that of "no-self" (anatta). The former notion implies that everything is in flux, nothing is permanent or static. "The world is in continuous flux and is impermanent," the Buddha told his disciple Ratthapala. In the Anguttara-nikaya there is the passage: "Oh, Brahmana, it is just like a mountain river, flowing far and swift, taking everything along with it, there is no moment, no instant, no second when it stops flowing, but it goes on flowing and continuing. So Brahmana, is human life, like a mountain river" (Rahula 1959, 25–26).

Many writers have noted the affinities between Buddha's stress on process and change and the philosophies of Heraclitus and Hegel. The pre-Buddhist philosophy of Kapila was a materialistic philosophy that also stressed that the world (prakrti) was a dynamic, self-becoming process and it is of interest that two of Buddha's teachers were Samkhya philosophers. But Samkhya was a dualistic philosophy which bifurcated spirit (purusa) and matter. There is a sense in which early Buddhism took over the Samkhya doctrines, but discarded the permanent and unchanging self (purusa, atman).

For the Buddha, charge and impermanence is the essential characteristic of all reality (Murti 1955, 62). The Buddha, as with Nagarjuna in a later period, was essentially a process or dialectical philosopher, in search of a psychological state that would give a person a sense of peace and tranquillity in an otherwise turbulent and changing world.

This brings us to the concept of no-self (anatta, Skanatma), which is somewhat unique to Buddhism, although there are echoes in both Taoism and Hume's philosophy. I have discussed the doctrine of "no-self" (soul) elsewhere,[10] but some brief comments may be made. Firstly, the doctrine can only be understood if set within the context of Hinduism, where the notion of the soul/self (atman) was taken to be an immutable eternal substance or entity within each person and a manifestation of the absolute spirit (Brahman). This identification of the self with the divine absolute went hand-in-hand with a denial of the reality of the world as it appeared (seen as maya, illusion). Salvation, in the Hindu context, consisted in removing this evil of ignorance and realising the oneness of the eternal atman (self) with Brahman, the union (moksha), being achieved through knowledge (or gnosis) (jnana), action (karma), devotion and ritual (bhakti), or physical discipline and meditative states (yoga).

Buddha repudiated the concept of atman and all that it implied, whether in terms of an immortal soul, or in terms of a creator/absolute spirit. Thus although in Buddhism there is a notion of rebirth, of the continuity of causes and conditions, there is no concept of reincarnation, on the transformation of immortal souls (Saddhatissa, 1971, 41). The tendency of many Western Buddhist scholars to interpret salvation (nirvana) as implying some kind of union of the self with a transpersonal ultimate reality—described as a "universal mind" or "all self"—seems to me highly misleading. It is also problematic to equate the Buddhist concept of the "void" (sunyata) with the Christian godhead or the Hindu Brahman.[11] There is also a common tendency among Western scholars to interpret Buddhist teachings as implying some form of subjective idealism. It is thus suggested that the Buddha saw the world, like the Hindu scholastics, as an illusion (maya), or that the mind is the "first cause" and thus that reality is "mind-only," or mind-independent.[12] Although subjective idealism is an important school of philosophy within Buddhism, for the common doctrine of the Yogacarins was that of Citta-Matra—all that exists is mind or consciousness—most Buddhists tend to follow the philosophy of Nagarjuna (circa 150 AD), known as Madhyamaka (the middle way). Nagarjuna affirmed that the natural world has no inherent static existence "out there," with each thing having an immutable essence or soul (eternalism). But this did not imply that the external world has no existence or reality and is merely a projection or creation of the mind (nihilism). Nagarjuna's concept of "no-self," sunyata (void or emptiness), suggested that things do not have a "soul" (essence) or independent existence; it does not imply that natural phenomena do not exist at all—and all is "mind" (Tenzin Gyatso 1995, 41–46).

For the Buddha (and Nagarjuna) the external world has a reality, but this reality (including humans) is a process, inherently changing, nothing having any permanence. Thus the Buddha also implied that the empirical self (as distinct from the soul) had no permanency or inherent reality. The human person was conceived, in Buddhist psychology, as consisting empirically of five categories of phenomena, termed aggregates (skandhas).

These are matter or corporeality (rupa), sensation or feeling (vedana), perception (sanna), will or action (samskara), and consciousness (vinnan). For the Buddha, all these aspects of human life are constantly changing and there is no entity on self that can be described as having any permanency, other than as a convenient name or label for the unity of these five aggregates. There is no persistent self, for "life is a constant flux." Buddha likened the human self to a chariot, that could be known only from its various parts and that had no essential unity. What the self of mine is permanent, stable, eternal, Buddha remarked, is pure speculation (Conze 1954, 74; 1959, 148–49). But again, the Buddha was not suggesting a form of nihilism, the complete "annihilation" of the self—this would make the idea of salvation incomprehensible—the truth is the "middle way."

Recognising the ubiquity of suffering, which is linked to the three "fires"—hatred (anger), greed (craving), ignorance (delusion)—and the "dependent origination" of all existents, the Buddha suggested a path of deliverance, a way of salvation. It implied a middle way, between on the one hand, a life given over to sensual pleasure and on the other, self-mortification and various forms of asceticism, which denies the senses and the body. The famous eightfold path consists essentially of three aspects: morality, meditation, and wisdom.

The first, ethical conduct (sila) consists of five basic precepts, namely to refrain from taking life, stealing, misusing the senses, telling lies, and becoming self-intoxicated through the taking of alcohol or drugs. The aim is to avoid any action that might harm others in any way and it involves compassion (karuna) and love for the world and all living creatures—"sentient beings." It advocates attitudes and actions that promote love, harmony, and happiness—and Buddhism is strongly opposed to any kind of war or the involvement in any kind of activity that entails suffering and harm to others. It also implies abstention from slanderous, idle, and dishonest speech. Tenzin Gyatso, the fourteenth Dalai Lama, stressed that morality is the foundation of the Buddhist path.

The second aspect is meditation (samadhi), or training in higher concentration. Samadhi has been defined as "contemplation on reality." In

Buddhism there are many forms of meditation, all of which are designed to increase awareness and to counter the "poisons" of craving, anger, and ignorance. The emphasis is towards concentration and insight (dhyana). It is important to note that the Buddha emphasised mindfulness, awareness, meditation that is contemplation and leads to insight: he does not advocate self-awareness, or going into trance states. The emphasis on insight is deemed to be quintessentially Buddhist (Rahula, 1959, 68; Gombrich 1988, 64). One well-known Buddhist scholar suggests that the aim of meditation is to "generate genuine insight into the ultimate nature of reality" (Tenzin Gyatso 1995, 20).[13]

The third aspect of the path to enlightenment is wisdom (panna), which demands a radically new way of looking at the world. It is described as the "ultimate and main element of the path" (Lamotte, 1984, 53). Wisdom, or gnosis, entails the cultivation of a sense of detachment and the understanding of the true nature of reality, as embodied in the four noble truths. It implies an attempt, as Edward Conze put it, "to penetrate to the actual reality of things as they are in themselves" (1959, 145).

This is not knowledge in the ordinary sense, but a kind of intuitive understanding—Sangharakshita describes it as an "unmediated spiritual vision" (1980, 16)—that sees all things directly, vividly, and truly and is free of misconceptions and prejudices. It is a philosophical outlook that has affinities with Spinoza and his notion of *scientia intuitiva* and Husserl's phenomenology.

The state of being known as nirvana, enlightenment, is described by the Buddha, often using images, in the following ways: It is like a lotus flower rising over of the water; it is like a mountain peak where seeds (passions) cannot grow; it is like space, or the wind, something unconditioned, but which can be experienced by the mind; it is a state of "pure consciousness" which entails freedom from becoming (the five aggregates) and thus the end of suffering; it is an awareness of things "as they really are."[14]

To what extent this path involves the assumption of a monastic life is none too clear, but Rahula suggests that the attainment of nirvana can be realised by men and women living ordinary lives (1959, 77; see also Carrithers 1983, 71–73). As indicated above, many Western scholars give a religious gloss to the concept of nirvana, seeing it as a form of religious mysticism. But as Rahula stresses nirvana is a synonym of truth; it is a state of psychological awareness that is lived and experienced. It does not entail the annihilation of the self because there is no self to annihilate. It does not involve the merging or fusion of the individual self with some absolute spirit or "universal self" because the Buddha considered such metaphysical ideas illusory. But the Buddha did insist that the attainment of nirvana gave the individual a sense of freedom, equanimity, and joy, and that it could be attained in this life, or at least, it entailed no rebirth when this life comes to an end (Gombrich 1984, 9).

2. The Limitations of Buddhism

Although canonical Buddhism advocates the "middle way," the social implications of the Buddhist approach to suffering is a radical detachment from the material world, and a repudiation of everything that constitutes or attracts the empirical self. The Buddha was positively hostile towards the enjoyment of the senses, especially relating to food and sex, and taking "delight in the senses and their objects" was inevitably seen as leading to desire, grasping, attachment, and thus to suffering (Conze 1959, 156).

Buddhism, in fact, with its emphasis on consciousness, not practical activity, on "detachment" and the "extinction" of desires, on being a "refuge" and an "island" to oneself, presents, it seems, a very one-sided view of human subjectivity. It is hardly surprising, then, that the Sangha, the order of Buddhist monks—which along with the Buddha and the dharma is one of the "three jewels" of Buddhism, and thus forms a central focus of Buddhist religion—should advocate celibacy and a vegetarian diet, and repudiate any active involvement in the material world, even agricultural production. As Gombrich writes, Buddhism is not concerned with God or the world; it is focused on suffering humanity, on morality, meditation, and gnosis. Buddha's teachings are not concerned with shaping human life in the world, but rather focused on liberation, on "deliverance," on "release" from the world and its inherent suffering. "The Buddha urged those who wished to escape from suffering to follow his example and renounce the world" (1984, 9). The care and the upbringing of children and the production of food and the basic necessities of life were left to others, mainly women and local communities.

It is however possible to go beyond the greed, conflict and "egoism" that Buddha saw as a fetter and a delusion, without this involving detachment from the natural world, the body or society. These could and should to be embraced in a naturally satisfying manner, for detachment from the world is not an existential possibility except in thought. Starhawk (Miriam Simos) repudiates the notion that "all life is suffering" and declares it to be a thing of wonder. Writing as a theist she declares that "escape" from samsara, the "wheel of birth and death," is not a viable option (1979, 42).[15]

Although Buddhism can be viewed as a form of radical humanism, it is not anthropocentric, for compassion, "love without attachment," was not restricted to humans but extended to all living beings. This has led many Western Buddhists to emphasise the notion that Buddhism is an "ecological religion," a "cosmic ecology" (Bachelor and Brown, 1992, viii). But Western Buddhists tend to interpret Buddhism as a religious system that constitutes a Dharma Gaia, so that it accords with Western psychological, religions, and ecological conceptions.[16] A number of misleading interpretations may be noted about such eco-Buddhism. Firstly, as earlier indicated, Buddhism tends to be interpreted as a form of subjective idealism. It is thus suggested that in reality no distinction can be made between subject and object, or between things in the world, such that the "universe is a seamless undivided whole" (Batchelor, 1992, 33). This is somewhat misleading. If this were indeed the case, it would make nonsense of the doctrine important in Buddhism that all things and phenomena are interconnected, interdependent, and conditioned. That Buddhism views "humanity as an integral part of nature" only makes sense if a distinction not dualistic can be made between humans and the world. The world for the Buddha is not "illusory," or an "undifferential unity," or mind-only (consciousness). What is illusory is a state of mind (ignorance), that fails to understand that everything in the world (events, things, mind, self) is conditioned, transitory, and in process. It is important then to distinguish as many Buddhists and postmodernists fail to do between the nature of reality (ontology) and forms of consciousness or knowledge (epistemology). Wisdom for the Buddhist is also a state of mind, but it is not so much cognitive, as an experiential state of "pure consciousness" that comes from non-attachment. It is this state of mind that is "undifferentiated" and for the Buddha blissful and ecstatic. But it entails detachment, not engagement with the material world. As one Tibetan monk of the eleventh century expresses it through a vision: "If you have attachment to this life, you are not a religious person. If you have attachment to the world of existence you do not have renunciation. If you have attachment to your own purpose, you do not have the enlightenment thought" (Batchelor, 1987, 212). The Buddha form of salvation thus advocates a radical detachment from the material world.

There is the famous story that when during a famine the local villagers refused to give food to the monks, one of the followers suggested that they should take up agriculture, for as the earth is "rich and as sweet as honey it would be good if I turned the earth over." But the Buddha refused and rebuked the monk, as agriculture would harm the creatures of the earth (Batchelor and Brown 1992, 13). This may be seen as expressing a caring attitude towards living beings (insects, worms) but it also affirms the radical detachment from—not engagement with—the natural world that the Buddha advocates. Other less enlightened humans were left to produce the food that the Buddhist monks needed to sustain life. Some Buddhists have even suggested that there should be no intention in human life to live "at the expense of any other creature," and to follow the precept "non-injury to life" is deemed to be the minimal code for every Buddhist" (de Silva 1992, 23). But as we are intrinsically organic beings, human life is not possible without the taking of life, and thus other living beings. The Buddha however (like vegetarians) tended to put on emphasis only on sentient "beings" (animals), and to completely ignore the fact that plants also are living beings.[17] Humans are not like bees, who can obtain wealth like a bee collects nectar from a flower, harming neither the fragrance nor the beauty of the flower. They have to labour, that is interact with nature, and like other mammals their livelihood is always at the expense of other living creatures.[18]

Secondly, the Buddhist doctrine of "no-soul" (anatta) is usually interpreted by eco-Buddhists as simply implying a rejection of the Cartesian subject—the self as an independent, autonomous, egocentric being, rather than a refutation of the "soul" and the empirical self. Nobody in their social praxis has even conceived of the self as radically separate or autonomous, for we all recognise the need to eat and breathe and acknowledge our social existence. But interpreting the self in Cartesian fashion allows Western Buddhists to interpret Buddhism in Jungian terms as a kind of "transpersonal psychology." They thus come to advocate the development of an "ecological self" (Macy, 1990). Whether this accords with the radical nature of the Buddhist doctrine of "no-self" (anatta) is, of course highly debatable. As Timmerman suggests, for the Buddha it was the self and not the natural world that was the problem (1992, 69).[19]

Thirdly, it is I think misleading, as noted earlier, to equate the Buddhist doctrine of the void and nirvana, with the Hindu conception of Brahman—as did many nineteenth century romantics and many Western Buddhist scholars, such as Conze and Humphreys.[20] Nirvana describes a psychological state of "pure consciousness"; it is not an ontological reality as is the Hindu brahman, the Christian God, the Islamic Allah or the neo-platonic one—however much these mystical traditions may share the idea of "gnosis"—mystical contemplation. Nor, unlike Taoism and Spinoza's pantheism, does Buddhism imply a "mystical identification with nature." Buddhist ontology, like Taoism, implies an ecological perspective, but as a religious practice it is profoundly antiecological in that it advocates a radical detachment from the world. There is some truth in Timmerman's suggestion that Buddhism is a kind of "mystical materialism" (1992, 69). But Buddhism does not suggest that the "sacred" or "divinity" can be discovered in nature, as this is reflected in the writings of those scholars and naturalists who were deeply influenced by the transcendentalist (idealist) tradition—Thoreau, Muir, and Aldo Leopold. Buddhism is neither a form of animism nor a spiritualist monism, as expressed in the religious philosophies of Advaita Vedanta and Neo-Platonism. Nirvana, for the Buddha, is a "state of the individual," attained through "nonattachment"; it is not an "entity or being" (Smart 1972, 23–24).[21]

Fourthly, it must be emphasised that many of the "ecological traditions" that are to be found in Buddhist countries like Ladakh and Thailand—organic agriculture, herbalism, a sacred attitude to landscape, especially trees, soft-energy systems, economic systems based on sharing—are not specifically Buddhist. Indeed, have little to do with Buddhism. They are part of the organic traditions of the peasantry—aspects of which (like animism and agriculture) were repudiated by the Buddha. They are in fact, pre-Buddhist. Helena Norberg-Hodge's (1992) panegyric on the ecological traditions of Ladakh (an old Himalayan Kingdom, now a part of India) not only ignores the fact that half the population of Ladakh are Muslims, but misleadingly conflates Buddhism with the ecocentric traditions of peasant communities. Such traditions are pre-Buddhist, and Buddhism is parasitic upon them, even though Norberg-Hodge emphasises the benign influence of the monastic elite. Such traditions, of course, are found throughout the world, in the Andes as well as in the Himalayas and among people who are not Buddhists.

It expresses an ethos and way of life that Ladakhi peasants share with European and Andean peasants (who are nominally Christians) and the hill people of nearby Uttar Pradesh (who are nominally Hindus).

It has often been suggested that Buddhism is a pessimistic philosophy, that it is a "world-denying" religion. Buddhism does not suggest a rejection or repudiation of the world, but rather detachment from it. Whether such a "lack of attachment" then allows one to enjoy nature as an aesthetic experience, as many eco-Buddhists suggest, it is difficult to say. Lily de Silva writes, "Buddha and his disciples regarded natural beauty as a source of great joy and aesthetic satisfaction" (1992, 27), and there is no doubt that many Buddhists throughout history have expressed positive attitudes towards nature. The essence of compassion implies respect for sentient beings, as well as for the earth itself. One of the most famous Buddhists who expressed a nature-ethic was the Tibet monk Milarepa (1052–1135). A poet and a solitary, Milarepa is one of the most esteemed monks in Tibetan history and his songs are well known. The "natural existence of the phenomenal world," he sang, were his "books" through which he sought understanding and enlightenment But his essential aim was enlightenment, to become detached from the world, not to engage with the world, or to enjoy it—for delight could easily lead to attachment. In the scriptures it records that Milarepa sang songs for local hunters, extolling the virtues of the mountains and living creatures, in exchange for meat and barley which he gladly cooked and ate. Milarepa appears not to have been a vegetarian (Batchelor, 1987, 109; Batchelor, 1992, 13–14).[22] Other Buddhists have been more negative towards the natural world, stressing the importance of liberation from a "world of darkness" (O'Neill, 1995, 195).

Many scholars, in fact, have noted the similarities between Buddhism and gnosticism. It is important to note, however, that these two religious traditions are completely dissimilar in terms of their underlying ontology. What they have in common is the emphasis on gnosis, spiritual enlightenment through meditation and esoteric knowledge. This is clearly brought out in Edward Conze's famous essay "Buddhism and Gnosis," originally published in 1930. The basic similarities include: salvation through gnosis (Jnana, contemplating knowledge); an elitist emphasis between the spiritually awakened (Peifecti, Aryas) and the ordinary people; a predilection for the esoteric and a mystical monism that puts an emphasis on light and the power of meditation states and sacred mantras or formulas. Conze, however, affirms that both Buddhism and gnosticism share "a yearning for union with the one" (1995, 18), without recognising their very contrasting metaphysics. For Buddhism, as earlier discussed, is a form of mystical naturalism that repudiates both the "soul" and the "divinity" as illusions. In contrast, as a spiritual monism, gnosticism affirms both.[23] More important Buddhism advocates detachment from the world as a site of suffering, while gnosticism repudiates the natural world entirely as the evil creation of the demiurge. Needless to say, neither gnosticism nor Buddhism are nihilistic, for while both stress the suffering and inherent "unsatisfactory" nature of worldly existence, they both offer "salvation" from this world of suffering and thus give meaning to human life.

I have stressed above that Buddhism, as it was expressed by the early Theravadins, suggested a path of spiritual enlightenment that entails a sense of detachment from everyday existence. This implies giving up sex, not eating meat (for animals embody life's most vibrant form), and not engaging in agriculture. Thus rather than indicating an ecological perspective (in its practices) or being an ecological lifestyle (as most of its Western adherent claim) Buddhism from its inception was anti-ecological, or at least antithetical to the organic traditions of the peasantry. This may be justified by considering the following:

1. Buddhism denigrated the shamanistic way of life of early human communities, where the hunter expressed his affinity and reverence for the animal he hunted It was hostile, too, towards animistic beliefs, in which natural phenomena, mountains, animals, trees, were seen as embodying spiritual beings or powers. In Tibet where the entire landscape is pervaded by associations with local deities and spirits, Buddhism is explicitly concerned with controlling and subduing these spiritual agencies. Although in Tibet a form of shamanistic Buddhism developed (Samuel 1993), clerical Buddhism has historically always been antagonistic towards and has even attempted to

suppress the spirit cults. With regard to Siberian shamanism, Ronald Hutton has noted that in terms of aggression and bigotry there was not much to choose between Islam, Christianity, and Buddhism, "for Buddhist monks wrecked pagan shrines and denounced shamanism with the same energy as adherents of the other two faiths" (1993, 17).

2. Buddhism lauded meditative states in remote forests or closed single-sex monasteries, away from both subsistence agriculture (on which it is parasitic) and the bustle of family life and children. This is one of the reasons, I suppose, that Buddhism always had a strong appeal to the deep ecologists, who appear to envisage nothing between an anthropocentric and exploitative attitude towards nature and the aesthetic enjoyment of nature in wilderness regions. Our dependence on a productive relationship with nature is ignored entirely by both the Buddhists and the deep ecologists.[24]

3. Buddhism preached the path of Nirvana, non-attachment, the "extinction" of desire. Of personality, rather than suggesting a sense of self which develops in a reciprocal relationship with both the natural world and other humans. It appears again that the Buddha saw nothing between the egocentrism, the greed, the hedonism, and the ethic of dominance and power of the caste to which he belonged—the *Kshatsiya*[25]—and complete detachment from the world, including that of the family and the social and natural worlds of the peasant community. Of course, as we have continually stressed, complete non-attachment is not a living option and even the most spiritual of gnostic mystics had to eat locusts and wild honey, or like Milarepa, nettles, in order to survive.

Buddhism as a "religion" emerged in the third century BC when the Manrya King Ashoka adapted and developed Buddhism as the state ideology of his expanding empire.[26] It provided the state with a "binding factor" in a cultural sense, its nonviolent ethos suiting the rulers, as well as facilitating extensive trade connections through the monasteries. Buddhism has thus historically always been associated with the state, with men, with a literary tradition and it has been distinct from the organic agricultural traditions of the peasantry. To conflate Buddhism with organic agriculture, as does Norberg-Hodge, is highly misleading. No true (male) disciple of the Buddha—if he followed explicitly his teachings—would soil his hands in agricultural work (better to meditate in the forests), or have sexual relationships with a woman, or eat the flesh of animals. Peasants do all of these things. As a European peasant said when he was hauled before the Inquisition: "I am no heretic, for I have a wife and lie with her and have children and eat flesh." He would not have made an enlightened Buddhist.

Although it is somewhat misleading to equate the teachings of the Buddha with the Buddhist religion as it developed over the centuries (in the same way one must distinguish the teachings of Jesus from the doctrines of the Christian Church), nevertheless, incipient within the teachings of the Buddha is a hierarchical conception of the social order, with respect to both gender and "class." Buddha's views on women are well known—although writers like Sangharakshita try to defend him—for the Buddha explicitly described them as the inferior sex, as well as seeing them as a source of anger, passion, and sexuality that would inevitably lead men away from the path of enlightenment.

Mary Mellor (1992, 45) notes that while Schumacher favours Buddhism as the basis of a green spirituality, it is quite explicit about the inferiority of women. The fact that Buddhism has a benign, a detached attitude towards nature is no guarantee, she writes, that it has a benign attitude towards women.[27]

We have already noted that early Buddhism was antithetical to agriculture, and has thus always been exploitative of the peasant producers—and where Buddhism has developed the monks have always constituted a ruling exile. In places like Tibet the monasteries had a monopoly of literacy—and precious few people outside the ruling exile could read and write—and were the centres of pomp, wealth, and power. They owned large acreages of land (usually in the most fertile valleys), worked of course by the peasants. Buddhist states often had much power, controlling trade (through the monasteries), and monks often formed their own military contingents—in spite of the Buddha's injunction against violence. Early Buddhist states in Tibet, as elsewhere, periodically had campaigns, as we have noted, to suppress local shamanistic cults. Tibet, before the Tibetan invasion, consisted of independent theocratic states, with essentially feudal relationships between the peasants and the monks and nobles.

Although we must deplore the military occupation of Tibet by Chinese troops and armed police, and the violation of human rights that still continues there and although one should support the right of Tibetan people to govern themselves as they wish, this ought not to blind us to the realities of the Buddhist state. Geoffrey Samuel stresses that the Tibetan states had limited coercive powers, only to highlight the fact that in other Buddhist countries—Sri Lanka, Thailand, Burma—a symbiotic relationship existed between Buddhism and these Asian states. For the Sangha, the community of monks was under state control, indeed was the ideological wing of these Buddhist states and the suppression of the spirit cults or non-orthodox forms of Buddhism was a constant theme (Samuel 1993, 27–29).

Moreover, when one observes contemporary Buddhism in practice, such as at the Jokhang temple in Lhasa, one realises how far it has diverged from Buddha's original teaching. For there is an emphasis on faith, on ritual oblations, on penance, on prostrating oneself before the images of the "Buddha" and his various incarnations and on the authority of the scriptures. The knowledge that Western Buddhists praise so highly is largely encapsulated in chants and scriptures, and although these are insightful and important regarding moral and psychological issues, they are little concerned with trying to understand the social and natural world in which we live. Such knowledge the Buddha always contemptuously dismissed as unfruitful in the quest for salvation.

In the above paragraphs I have tended to paint a less rosy picture of Buddhism than that normally portrayed by its Western adherents. But this is not to deny that Buddhism has a lot to offer anarchism. In its emphasis on nonviolence, in expressing compassion and sympathy for all living creatures, and in its ethical code, the Buddha's teachings still have a contemporary relevance. The Buddha stressed that generosity should replace greed, that compassion ought to take priority over hatred and contempt, and that wisdom—as a realistic awareness of the nature of phenomenal existence—is better than following delusions. And the delusions of which he spoke were those of religion—the belief in an all-powerful God, in the spirit world, in an immortal soul that survived death.

Throughout the history of Buddhism individual monks—like Milarepa—have expressed in their writings and poetry an ecological perspective in their compassion for the natural world, in ways that are reminiscent of the early romantics in Europe—who also belonged to the literati. Many have seen the Sangha, the Buddhist order of monks, as an early form of communism, a democratic community which had renounced private property. But as Debiprasad Chattopadhyaya (1959) argued, the Sangha was simply an emulation of a clan-based society—apart from the fact that it excluded women and children and was under the absolute authority of the abbot. The Western Buddhist order still advocates single-sex fraternities, seeing the association of people with the opposite sex as sullying the purity of the Buddhist path to salvation.

It is perhaps appropriate that I should conclude with a quotation from the Buddha. "Do not be misled by proficiency in the scriptures, nor by mere logic and inference, nor after considering reasons, nor because the recluse is your teacher. But when you know for yourselves: These things are not good, these things are faulty, these things are contradicted by reason, these things when performed lead to loss and sorrow—then do you reject them" (Humphreys 1987, 71).

Notes

1   See for example Badiner 1990; Batchelor and Brown 1992; and Marshall 1992, 41–53.

2   Given its long history and the fact that Buddhism is found among a wide range of different people and cultures, there are, as the saying goes, more than 57 varieties of Buddhism. I focus the discussion here mainly on the canonical or clerical Buddhism of early Theravadin societies. For useful general studies of Buddhism and its history see Ling 1973, Bechert and Gombrich 1984.

3   Saddhatissa 1971, 19. For a good short introduction to Buddha's life see Carrithers 1983.

4   Conze 1954, 87. The text reads: "Monks, I will teach you the Dharma—the parable of the raft—for crossing over, not for retaining."

5   Dennis Lingwood (Sangharakshita) was born in 1925 in South London of working class parents. He spent more than a decade at Kalimpong in the Himalayas and became a Buddhist monk. In 1967 he founded the Friends of the Western Buddhist Order, and has since then written numerous books—both scholarly and readable—on the Buddhist tradition. An account of his work is given by Subhuti (1994).

6   Humphreys 1987, 93–94.

7   Rahula 1959, 51–52. Walpola Rahula's text *What the Buddha Taught* is considered one of the best short introductions to Buddhism. Rahula was a controversial scholar-monk from Sri. Lanka, who stressed that Buddhism entailed a radical politics. However such politics have been interpreted by other Buddhists as implying too close a link between Buddhism and Sinhalese ethnic-nationalism (see Sangharakshita 1986, 69–91).

8   Anthropologists have long recognised that among Buddhist communities of South and Southeast Asia, Buddhism often takes several forms and that most people are committed towards worldly concerns not enlightenment. Three general spheres of religious activity have been described, a practical orientation, which often involves spirit cults and shamanic practices; a Karma orientation that is concerned with the ideology of merit and attaining a better rebirth and finally, a Bodhi orientation which involves the pursuit of enlightenment, seen as an "escape" from the cycle of rebirth and thus worldly concerns canonical Buddhism emphasises this Bodhi, salvation, orientation (see Spiro 1971, Southwold 1983, Samuel 1993 for excellent studies of Burma, Sri Lanka, and Tibet respectively).

9   John Snelling, significantly, implies that it has both ontological grounding, suffering (dukkha) being the "flawed nature of all that exists" (1987, 64) and that it is derived from our distorted "perceptions" of the world (85).

10   See my *Anthropology of the Self*, 1994, 57–69, and my discussion of Samkhya philosophy, one of the six schools of Indian philosophy, pages 73–75.

11   For examples of this tendency see Owens 1975, 165–75; Humphreys 1987, 15–19; Snelling 1987, 8. Important critiques of this tendency, which conflates the metaphysical doctrines of quite distinctive religious (mystical) traditions, are to be found in Katz 1983.

12   See, for example, Marshall 1992, 42; Batchelor and Brown 1992.

13   Important text on Buddhist meditation include Nyanaponika Thera 1962, Sangharakshita 1980.

14   These extracts are taken from Conze 1959, 156–59; Lamotte 1984, 51–52; Sangharakshita 1990, 209.

15   Of American Jewish background, Starhawk is one of the leading gurus of feminist witchcraft in the United States. She is an advocate of goddess spirituality, animism, and spirit reincarnation, so, contradicting what she says, allows, after the dissolution of the body of death, the "spirit to prepare for a new life." Buddha thought such an idea an "illusion." Starhawk's electric spiritualism seems even less compatible with anarchism than is Buddhism.

16 See the studies of Ian Harris (1991, 1995) on this issue. Harris notes that the eco-Buddhism of many Western Buddhists represents a substantial shift away from traditional Buddhist cosmology and practices. He notes in particular how the Buddhist understanding of reality as indicated in the canonical texts, has been replaced by Western metaphysical notions drawn from process theology and the Christian stewardship tradition. Needless to say the adherents of every form of religion—paganism (animism, shamanism), Judaism, Christianity, Islam, Hinduism—now claim that their religion entails an environmental ethic which emphasizes harmony between humans and nature.

17 The Buddha, of course, emphasising both compassion and non-attachment, also stressed that monks should refrain from harming plant life; see Harris 1991, 107–9.

18 A recognition of this fact, and the interdependence of all living beings and the world, is what constitutes an ecological outlook. The Buddhist emphasis on nonattachment denies this dependence and the intrinsic organic and inorganic links between humans and the material world—though this detachment can only be psychological.

19 I have discussed the limitations of interpreting Buddhism as a "psychology of self-realisation" elsewhere (1994, 66–67).

20 See note 11 above.

21 Succinct outlines of the Vedanta tradition in Hinduism (Sankara) and neo-Platonism (Plotinus) are given in Peter Marshall's *Nature's Web*, 1992, 31–34, 84–87. On transcendentalism and ecology see Worster 1997.

22 It has to be noted that there is no intrinsic connection between Buddhism and vegetarianism. Buddha was a meat-eater, and as a coherent ethical doctrine vegetarianism was a late development in both Hinduism and Buddhism (Harris 1991, 106).

23 For important studies of gnosticism see Jonas 1958 and Pagels 1982.

24 It is of interest that one deep ecologist Dolores LaChapelle (1988) makes no mention of Buddhism, but like Starhawk combines pantheism (Spinoza, Taoism), animism, and theism in an eclectic mix and advocates ritual ceremonial.

25 The Kshatriya, in early Hindu society, were a caste of aristocratic warriors, who, in contrast to the *Brahmin* (priests), *Vaisyas* (merchants), and *Shudras* (service castes), were responsible for upholding the *dharma* (moral law). From this caste, the rulers of the petty states were drawn. Buddha belonged to this aristocratic ruling caste.

26 The Maurya dynasty, under Chandragupta, Ashoka's grandfather, united much of India as an imperial domain around 324 BC. See the classic history by Romila Thapar (1963).

27 On Buddhist attitudes to women see Ling 1981, 193; and Subhuti 1994, 162–75. Sangharakshita has consistently expressed views that are anti-feminist and seems to regard sexual relationships and attachments as antithetical to the "spiritual life."

References

Badiner, Allan Hunt, ed. 1990. *Dharma Gaia: A Harvest of Essays in Buddhism and Ecology*. Berkeley: Parallax Press.

Batchelor, Martine, and Kerry Brown, eds. 1992. *Buddhism and Ecology*. London: Cassell.

Batchelor, Stephen, ed. 1987. *The Jewel in the Lotus: A Guide to the Buddhist Traditions of Tibet*. London: Wisdom.

————. 1992. "The Sands of the Ganges." In *Buddhism and Ecology*, edited by Martine Batchelor and Kerry Brown, 31–39. London: Cassell.

Bechert, Heinz, and Richard Gombrich, eds. 1984. *The World of Buddhism*. London: Thames and Hudson.

Carrithers, Michael. 1983. *The Buddha*. New York: Oxford University Press.

Chattopadhyaya, Debiprasad. 1959. *Lokayata: A Study in Ancient Indian Materialism*. New Delhi: People's Publication House.

Conze, Edward. 1951. *Buddhism: Its Essence and Development*. Oxford: Cassirer.

_____, ed. 1954. *Buddhist Texts through the Ages*. Oxford: Cassirer.

_____, ed. 1959. *Buddhist Scripture*. Harmondsworth: Penguin.

_____. 1995. "Buddhism and Gnosis." In *The Allure of Gnosticism: The Gnostic Experience in Jungian Philosophy and Contemporary Culture*, edited by Robert A. Segal, 173–89. Chicago: Open Court.

De Silva, Lily. 1992. "The Hills Wherein My Soul Delights." In *Buddhism and Ecology*, edited by Martine Batchelor and Kerry Brown, 18–30. London: Cassell.

Gombrich, Richard. 1984. "Introduction: The Buddhist Way." In *The World of Buddhism*, edited by Heinz Bechert and Richard Gombrich, 9–14. London: Thames and Hudson.

_____. 1988. *Theravada Buddhism*. London: Routledge and Kegan Paul.

Harris Ian. 1991. "How Environmentalist Is Buddhism" in *Religion* 21, pages 101–14.

_____. 1995. "Buddhist Environmental Ethics and Detraditionalisation." *Religion* 25: 199–211.

Humphreys, Christmas. 1987. *The Wisdom of Buddhism*. New Delhi: Promilla.

Hutton, Ronald. 1993. *The Shamans of Siberia*. Glastonbury: Isle of Avalon Press.

Jonas, Hans. 1958. *The Gnostic Religion*. Boston: Beacon Press.

Katz, Steven T. 1983. *Mysticism and Religious Traditions*. London: Oxford University Press.

LaChapelle, Dolores. 1988. *Sacred Land, Sacred Sex: The Rapture of the Deep*. Silverton, CO: Finn Hill Arts.

Lamotte, Etienne. 1984. "The Buddha, His Teachings and His Sangha." In *The World of Buddhism*, edited by Heinz Bechert and Richard Gombrich, 44–58. London: Thames and Hudson.

Ling, Trevor. 1973. *The Buddha: Buddhist Civilization in India and Ceylon*. London: Temple Smith.

_____, ed. 1981. *The Buddha's Philosophy of Man*. London: Dent.

Macy, Joanna. 1990. "The Greening of the Self." In *Dharma Gaia: A Harvest of Essays in Buddhism and Ecology*, edited by Allan Hunt Badiner, 53–56. Berkeley: Parallax Press.

Marshall, Peter, 1992. *Nature's Web: An Exploration of Ecological Thinking*. London: Simon & Schuster.

Mellor, Mary. 1992. *Breaking the Boundaries: Towards a Feminist Green Socialism*. London: Virago Press.

Morris, Brian. 1994. *Anthropology of the Self*. London: Pluto Press.

Murti, T.R.V., 1955. *The Central Philosophy of Buddhism*. London: Unwin.

Norberg-Hodge, Helena. 1992. "May a Hundred Plants Grow from One Seed." In *Buddhism and Ecology*, edited by Martine Batchelor and Kerry Brown, 41–54. London: Cassell.

Nyanaponika Thera. 1962, *The Heart of Buddhist Meditation*. London: Century.

O'Neill, Kenneth. 1995. "Parallels to Gnosticism in Pure Land Buddhism." In *The Allure of Gnosticism: The Gnostic Experience in Jungian Philosophy and Contemporary Culture*, edited by Robert A. Segal, 190–98. Chicago: Open Court.

Owens, Claire Myers. 1975. "Zen Buddhism." In *Transpersonal Psychologies*, edited by Charles T. Tart, 153–202. London: Routledge & Kegan Paul.

Pagels, Elaine. 1982. *The Gnostic Gospels*. Harmondsworth: Penguin.

Rahula, Walpola. 1959. *What the Buddha Taught*. London: G. Frazer.

Saddhatissa, Hammalawa. 1971. *The Buddha's Way*. London: Allen & Unwin.

Samuel, Geoffrey. 1993. *Civilized Shamans: Buddhism in Tibetan Societies*. Washington, DC: Smithsonian Institution Press.

Sangharakshita. 1980. *Human Enlightenment*. Glasgow: Windhorse Publications.

_____. 1986. *Alternative Traditions*. Glasgow: Windhorse Publications.

_____. 1990, *A Guide to the Buddhist Path*. Glasgow: Windhorse Publications.

Shaner, David Edward. 1989. "The Japanese Experience of Nature." In *Nature in Asian Traditions of Thought*, edited by J. Baird Callicott and Roger T. Ames, 163–81. Albany: State University New York Press.

Smart, Ninian. 1972. *The Concept of Worship*. London: MacMillan.

Snelling, John. 1987. *The Buddhist Handbook*. London: Century.

Southwold, Martin. 1983. *Buddhism in Life*. Manchester University Press.

Spiro, Melford E. 1971. *Buddhism and Society*. London: Allen & Unwin.

Starhawk. 1979. *The Spiral Dance*. New York: Harper Collins.

Subhuti, Dharmachari. 1994. *Sangharakshita: A New Voice in the Buddhist Tradition*. Birmingham: Windhorse Publications.

Tenzin Gyatso. 1995. *The World of Tibetan Buddhism*. Boston: Wisdom Publication.

Thapar, Romila. 1963. *Asoka and the Decline of the Mauryas*. London: Oxford University Press.

Timmerman, Peter. 1992. "It Is Dark Outside." In *Buddhism and Ecology*, edited by Martine Batchelor and Kerry Brown, 65–76. London: Cassell.

Worster, Donald. 1977. *Nature's Economy: A History of Ecological Ideas*. Cambridge University Press.

5

Capitalism: The Enemy of Nature (2003)

A review of Joel Kovel's *The Enemy of Nature: The End of Capitalism or the End of the World?* (London: Zed Books, 2002)

Joel Kovel's book *The Enemy of Nature* has been described as an "ecosocialist manifesto." It is written with zest and some eloquence and is full of useful insights. The feminist scholar Nancy Hartsock has suggested that the book is "highly original" and an impressive reworking of Marxist theory. It is indeed part of a movement that aims at the greening of Marxism. But its main thesis is hardly original, namely that capitalism is not only responsible for the present social crisis—growing economic inequalities, widespread poverty, political repression, human rights violations—but is also the "uncontrollable force" driving the environmental crisis. As Kovel puts it, capitalism is the "efficient cause" of the ecological crisis. This is hardly news to anarchists. Indeed Murray Bookchin, in his strident critiques of "deep ecology," has for several decades been stressing that the capitalist system is both inherently ecodestructive and unreformable—that it invokes the domination of nature and is the antithesis of a viable ecological society. Like Habermas, Bookchin also emphasised that contemporary capitalism penetrates into all aspects of our personal "life-world." Kovel's main thesis—that capitalism is the "enemy of nature"—is then hardly a new or original idea. But at the present juncture when the ruling consensus is that there is no viable alternative to capitalism, when radicals and liberals alike engage in identity or "lifestyle" politics—as Anthony Giddens describes them—and "leftism" and the "politics of emancipation" is declared to be old-fashioned or "redundant," Kovel's powerful ecological critique of capitalism is to be welcomed and applauded. It gives a succinct, substantive, and readable critique of the "monster that now bestrides the world"—global capitalism.

The book is in three parts. The first part gives an account of the ecological crisis—which manifests itself in various forms as global warming, deforestation, pollution, loss of species diversity—and of the nature of capitalism. Kovel offers a case study of the Bhopal catastrophe of 1984, when poisonous gas escaped from the Union Carbide factory—which manufactured pesticide—killing an estimated eight thousand people. Capitalism, that "gigantic machine for accumulation," as Kovel describes it, makes a fetish of technology, emphasises commodity-exchange at the expense of use values, and promotes a totally administered society—capital penetrating all aspects of the life-world of individuals and communities. The "unholy trinity" of the World Trade Organization, the World Bank, and the International Monetary Fund tend to fuse, as Kovel suggests, into an "iron triangle" of global accumulation, serving the needs of a "transnational bourgeoisie." Such globalisation does not imply the decline of the nation-state, for the United States has emerged as the hegemonic "enforcer" of the relentless expansion of capital. Indeed, Kovel suggests that there is a symbiotic relationship between corporate capitalism, national governments, and organised crime. He thus concludes that although capitalism has the fantastic ability to produce wealth, it also produces poverty, insecurity, eternal strife, and ecodestruction—the undermining of the life-support systems on which human life depends. He considers the casino capitalism of Las Vegas and the tendency of capitalism to create people with specific psychological dispositions—isolated, narcissistic, anxiety-ridden, calculating. For Kovel, then, capitalism is "anti-ecological," a premise that may be news to some Marxists but which is common knowledge to those in the socialist anarchist tradition.

In the second part Kovel offers some interesting reflections on ecology and on what Bookchin described as the "legacy of domination." With regard to the first topic Kovel outlines a philosophy of nature that Kropotkin (among others) had initiated at the end of the nineteenth century, and which may be described as evolutionary holism. Kovel gives a very cogent account of this ecological worldview, and stresses the following: that individuality and connectedness are both integral to human life; that humans have a nature and as species-beings have specific powers and capacities, of which subjectivity, imagination, and sociality are of crucial importance; that ecological thinking focuses not on an environment (as a set of things outside us with no essential structure) but on ecosystems or wholes, defined by internal relations; and that ecology is integrally tied to an evolutionary perspective. Kovel continually alludes to the need for a "spiritual life" or "spirituality" and quotes with seeming approval a saying of the Christian mystic Eckhart: "let us pray to God to be rid of 'God'" (253). But how the "spirit" (seemingly conceived as "god" or the "ultimate being," and which is institutionalised as religion) fits into an ecological philosophy is never explained. Kovel also emphasises the importance of dialectics, the recognition of "identity-in-difference" (139), and contrasts this "differentiated relationship" with the notion of "splitting," the psychological counterpart of what is usually described as "dualism." Kovel then offers a rather brief and somewhat conjectural historical account of the human "estrangement from nature" and the emergence of class society, the state, capitalism, and the "domination of nature" ethic. As a philosophical interlude he provides a useful critique of that "profound" thinker and doyen of the deep ecologists Martin Heidegger, questioning his usage of an undifferentiated "man." This seems to echo Bookchin's critique of deep ecology more generally.

The final part, "Towards Ecosocialism," addresses the question of "what is to be done." It begins by critiquing existing theories of political ecology—the reformist politics of the "New Democrats" like Al Gore, the notion that we can overcome the ecological crisis by some technological fix, the "green economics" stemming from neo-Smithian theorists (Jefferson, Schumacher, Korten) who essentially advocate a petty capitalism of independent producers, and the philosophies of deep ecology, bioregionalism, and ecofeminism. He also offers some cogent critiques of populism and Nazism—particularly emphasising the close affinities between neo-fascism and ecological thought as expressed in the writings of Heidegger, Bahro, and Gruhl. These critiques are not particularly novel, but they need to be emphasised and Kovel expresses them well and thoughtfully—particularly the critiques of the "wilderness" concept of the deep ecologists, and mother-goddess spirituality. Again much of this simply echoes Bookchin's earlier, harsh critiques of deep ecology and spiritual feminism.

But Kovel's criticisms of anarchism, specifically that of Bookchin's social ecology, regrettably, are typically those of many Marxists, in being trivial, misleading, or distorting—thus misrepresenting anarchism. One gets an inkling of this as soon as one reads the preface where Kovel writes: "Green politics tend to be populist or anarchist rather than socialist, hence Greens envision an ecologically sane future in which a suitably regulated capitalism . . . continues to regulate social production" (ix). This sets up a false opposition—a "splitting"—between anarchism and socialism, which for most anarchists is conceptually inept and historically invalid. Although, as at the end of the nineteenth century, there are individualist anarchists around who follow the likes of Stirner and Nietzsche, and some anarcho-primitivists (especially in the United States) who repudiate socialism as "leftism," the majority of anarchists have been socialists and have described themselves as libertarian socialists, or social anarchists or communist anarchists. Indeed Gustav Landauer's well-known text is titled *For Socialism*, and Kropotkin described anarchism as the most radical form of socialism. The implication that anarchism entails a "suitably regulated capitalism" is utterly misleading.

In specifically critiquing Bookchin's social ecology Kovel suggests that anarchists have a blanket condemnation of all forms of hierarchy and authority, and suggests that parent/child and teacher/student relations are, in a sense, positive forms of "hierarchy." This criticism is trivial. Bookchin used the term "hierarchy" to apply not to parental care or to pedagogic relationships but to institutionalised systems of domination, and no anarchist has ever repudiated all forms of authority. Didn't Bakunin famously reply to the question whether he rejected all authority: "Perish the thought. In the matter of boots, I defer to the authority of the bootmaker" (Dolgoff, *Bakunin on Anarchy*, 229)?

Then Kovel has the utter temerity to suggest that the anarchists, because of their hostility to Marxism, fail to recognise that the prime function of the state is to uphold capitalism, the class system, and thus they avoid giving central importance to the emancipation of labour. Thus Kovel suggests, with some pomposity, that Bookchin and the anarchists need to be moved along the "anti-capitalist" road (178, 189), notwithstanding the fact that anarchists have been at the forefront of recent anti-capitalist demonstrations in Seattle and elsewhere. However, a mere perusal of the writings of anarchists, whether anarchist communists or anarcho-syndicalists (from Kropotkin, Goldman, and Rocker to Chomsky and Bookchin) as well as an examination of anarchism as a historical social movements, will indicate that anarchism has always been "anti-capitalist" (and anti-religion) and has been concerned with the emancipation of working people from what they described as "wage slavery." Equally anarchists have always presented a powerful critique of Marxian socialism—its authoritarian politics and its support of "state capitalism." Kovel's critique of "actually existing socialisms" simply appropriates and replicates the anarchist critique of Marxist politics (historically exemplified by Bolshevism and Maoism) that goes back to the nineteenth century. Kovel then presents a distorted portrait of anarchism, which has always emphasised the symbiotic relationship between the modern state and capitalism, and has been both anti-state (libertarian) and anti-capitalist (socialist). It has always been concerned with the emancipation of labour, but through direct action rather than parliamentary politics, and, unlike Marx, was not exclusively focused on the industrial proletariat. Nor is Bookchin, and the anarchists more generally, "rigidly anti-Marxist" as Kovel contends. Anarchists, from Bakunin onwards, have always been appreciative of Marx's historical materialism and his critical analysis of capital, and have acknowledged his strident critiques—of religion as a form of ideology that masks inequalities, of market socialism (seemingly now espoused by both Kovel and John Clark), of bourgeois individualism, and of capitalism as an exploitative system. Bookchin, situating himself like Marx within the critical tradition of the Enlightenment—its legacy of freedom—has been strongly influenced by Marx. In fact, his *Ecology of Freedom* (1982) is permeated with Marx's ideas, for his whole philosophy represents a creative synthesis of Hegelian Marxism, ecology, and social anarchism. What Bookchin repudiates, in standing firmly in the anarchist tradition as a socialist movement, is Marx's statist politics.

This is precisely what Kovel has to offer in envisioning the transformation, the overthrow, of capitalism. For Kovel clearly sees the revolution as involving the "struggle for the state"—duly made democratic, with the revolutionary moment aiming at the "seizure of state power" (200). Such a state is seen as coexisting with a "free association of producers." Thus the "ecosocialist transformation" will involve the democratic state, party, or electoral politics (the party will be neither vanguard nor bourgeois but "green"—ecosocialist), the "state apparatus" in new hands (whose, one may ask, and for what purpose?), a market economy, committees with "state-like" functions, and prisons for anyone stepping out of line. His colleague John Clark has a similar vision of an ecocommunitarian realpolitik; namely, representative government with a coercive legal system, administrative bodies that dictate social policy, and a market economy! For Kovel, going "beyond" Bookchin implies an unholy marriage of Marxism and mysticism (spiritualism), a retreat from anarchism into the advocacy of some form of market socialism, replete with representative government and coercive institutions. It is not surprising therefore that dear old Bookchin, steadfastly defending the socialist anarchist tradition against Marxists, anarcho-primitivists, and mystical deep ecologists and feminists, quite apart from the liberal democrats, should be accused by Kovel of being "rigidly" anti-spiritual, "unrelentingly" dogmatic and sectarian, and "highly" Eurocentric. Spiritualism—of whatever variety (theism, polytheism, animism, or mysticism)—implies an ontology that is "rigidly" and dogmatically anti-secular and unecological. Defending a socialist revolution over reformism and stridently defending one's commitment to anarchism over reformism is no more "dogmatic" and "sectarian" than Kovel's plea for a new brand of Marxism—market socialism; and a refusal to abandon the insights of the Enlightenment secular tradition and to become besotted with either oriental mysticism or Maoism does not make Bookchin a Eurocentric pleb. It must be said, however, that Kovel's critique of social ecology is singularly free of the hostile vilification that mars the work of Bookchin's other critics. Perhaps Landauer was right: "Marxism is the professor who wants to rule"!

6

Down to Earth: A Tribute to Three Pioneer Ecologists (2006)

In the introduction to his Russell Memorial lecture in 1971, Noam Chomsky mentioned a Japanese farmer who had a wall poster which read: "Which road is the correct one, which is just? Is it the way of Confucius, of the Buddha, of Jesus Christ, Gandhi, Bertrand Russell? Or is it the way of Alexander the Great, Genghis Khan, Hitler, Mussolini, Napoleon, Tojo, President Johnson?" (2003, ix).

It is doubtful if Russell, a passionate sceptic and libertarian all his life, would have been entirely happy being associated with a reactionary like Confucius, or a religious mystic like the Buddha—but nonetheless this extract clearly and unambiguously depicts the false dilemma with which we are presented as soon as we begin to discuss ecology. Either we have to side with religious mystics and neo-pagans and cultivate a "sacramental" or spiritual attitude towards nature, or we are alleged to align ourselves with the positivist tradition and mechanistic philosophy, with aggressive imperialism, industrial capitalism, and agribusiness. In innumerable ecological texts the choice we are given is thus between either mechanism or spiritualism; either mammon (industrial capitalism) or god (religion); either Cartesian philosophy, with its dualistic metaphysics and its emphasis on the technological mastery of nature, or so-called "spiritual ecology" which embraces some form of religious metaphysics, either neopaganism, theism, pan(en)theism, or mysticism. For example, in a recent text by "spiritual ecologist" David Watson, which purports to offer a "deep ecological vision," the choice we are presented with is either the "prison-house of urban-industrial civilization" with its accompanying ideologies, or "primitivism"—entailing the wholesale rejection of technology, the affirmation of a hunter-gathering existence, and the embrace of neopaganism—tribal animism (Watson 1999).

What is lost in all this is that there is another ecological tradition that repudiates both mechanism *and* spiritualism, that while critiquing industrial capitalism and the megamachine, along with its anthropocentric and dualistic paradigm, does not go to the other extreme and embrace primitivism and some form of religious metaphysic. This is the tradition of organic or ecological humanism, a tradition that is particularly associated with three pioneer social ecologists: Lewis Mumford, René Dubos, and Murray Bookchin. In this article I shall briefly review some of the essential themes that emerge from their ecological writings.

Like many other scholars, these social ecologists stress that there is an essential "paradox" at the heart of human life, for there is an inherent duality in social existence, in that humans are an intrinsic *part* of nature, while at the same time, through our conscious experience and our human culture, we are also in a sense *separate* from nature. Mumford speaks of humans as living in "two worlds"—the natural world, and what all three scholars, following Cicero, call "second nature"—human social and symbolic life which is "within" first nature. Humans thus have a dual existence, in that they are simultaneously contemplative and active beings, both "constituting" (giving cultural meaning to) and being actively engaged in the natural world.

Fully embracing Darwin's evolutionary theory, Mumford, Dubos, and Bookchin emphasise that humans are a product of natural evolution, and that there is therefore no radical dichotomy between humans and the natural world, specifically other life-forms. All three scholars thus repudiate Cartesian philosophy, with its dualistic metaphysics—implying a radical dichotomy between humans and nature, and between the body and the mind—along with its atomistic epistemology, its anthropocentric ethic, which envisages the technological mastery of nature, and its mechanistic paradigm. Following Darwin, they emphasise that the world—nature—is not a machine but an evolutionary process, which can only be understood by an organic, developmental way of thinking and a holistic (relational) epistemology. This Bookchin describes as "dialectical naturalism." All three scholars therefore stress the crucial importance of historical understanding, especially with regard to biology. Dubos indeed affirms that nothing in biology makes sense except in the context of history. They are therefore critical of much social science which tends to emphasise a radical dichotomy between culture and nature, and, in over-emphasising the autonomy of culture, to even oblate biology entirely in the understanding of social life. Mumford, Dubos, and Bookchin are then all committed evolutionary naturalists.

But all three scholars are equally critical of all forms of reductive materialism, which tend to downplay the uniqueness of the human species, our "humanness," and they stress in particular the fundamental importance of human culture—technology, the arts, symbolism, philosophy, science—which make humans a unique species. Throughout their writings strident criticisms are therefore made of Social Darwinism, neo-Malthusian doctrines, behaviourism, and sociobiology. What they also stress is that humans, like all other organisms, do not simply adapt to environmental conditions but have creative agency, and Mumford indeed describes humans as the "unfinished animal," while Dubos, emphasising that humans are dialectically linked to nature, notes that human life tends to "transcend" its earthly origins. Yet their humanism also involves putting an equal stress on the autonomy and well-being of the human personality, and the development of an ethical naturalism that critiques both cultural relativism and religious absolutism, and emphasises a universal ethics that recognises the sociality and unity of humankind.

Remaining true to the Enlightenment tradition, they therefore emphasise the need to uphold its fundamental values, namely, liberty and the freedom of the individual, equality and social justice, cosmopolitanism and tolerance, and the need to develop a radical form of democracy. They acknowledge that there is a need to defend this tradition as against its neo-romantic detractors (see Bronner 2004). The key concepts of Mumford's, Dubos's, and Bookchin's evolutionary humanism are therefore wholeness, balance, diversity, autonomy, and mutualism. They particularly express the need to sustain both unity and diversity (personal, social, and ecological), both human subjectivity and social cooperation, both the flourishing of humans and that of the biosphere, its landscapes, and its life-forms.

Reacting against the Social Darwinian emphasis on conflict, struggle, and the "survival of the fittest," as well as against the atomism inherent in Cartesian philosophy and mechanistic science, Mumford, Dubos, and Bookchin all stress the importance of mutual aid and symbiosis in the understanding of the biosphere, as well as of human life. All three men thus warmly embrace and pay tribute to the ecological vision of the Russian anarchist-geographer Peter Kropotkin (see Morris 2004). This meant that they were not only critical of Cartesian dualism but also of the scientistic ethic, most famously developed by Francis Bacon in his *De Augmentis Scientiarum* of 1623, that views the natural world simply as a human resource and encourages human domination of nature.

There has, however, been a tendency among some scholars to suggest rather misleadingly, if not bizarrely, that "humanism" is simply a secular version of the Christian faith. That "humanism" posits a dualistic metaphysic that not only implies a radical "gulf" between humans and other life-forms, but suggests that humans have been given dominion over the earth, expressed in their technological mastery of nature. By such criteria

Mumford, Dubos, and Bookchin are not "humanists"! Scholars such as Ehrenfeld (1978) and Gray (2002) also offer as a rather misanthropic portrait of humans as being inherently destructive and predatory animals, and thus in need of salvation or redemption via some religious faith or mystique.

But Mumford, Dubos, and Bookchin, like prominent British humanists such as Julian Huxley, repudiate such misanthropic and anti-humanist sentiments, affirm the continuity of humans with the rest of nature, particularly other life-forms (embracing an evolutionary perspective), and not only deny that humans can be the "masters" of nature, but offer powerful critiques—long before Gray—of the Baconian ethic regarding the human "dream" of mastering nature. Such a Faustian attitude, Dubos argued, was not only misplaced and dangerous, but contrary to biology.

Both Mumford and Dubos were in many respects religious thinkers. Dubos is often considered to be an advocate of Christian stewardship regarding nature, though he actually proposed a *"scientific* theology of the earth." Mumford had a vague pantheistic sense of god, that was akin to that of Spinoza. But all three social ecologists when they spoke of "religion" or "spirituality" essentially implied a sense of wonder and respect towards natural phenomena, and the need to develop what Bookchin describes as an "ecological sensibility." They thus attempt to combine evolutionary naturalism with a form of humanism that is very different from that defined by Ehrenfeld and Gray, one that has been advocated by many contemporary philosophers. This form of humanism has the following characteristics: (1) it is naturalistic rather than supernaturalist, repudiating spiritualist explanations of natural and social phenomena, thus putting an emphasis on human reason, (2) it affirms the unity of humankind and a naturalistic ethics that recognises the existence of basic universal values, (3) it acknowledges the dignity of the human personality and the crucial importance of upholding such human values as equality, liberty, tolerance, and social solidarity, and (4) it suggests a relational epistemology that emphasises free inquiry and the importance of reason and science, as well as of the human imagination (Kurtz 1983, 39–47; Bunge 2001, 14–15).

What was crucial with regard to Mumford, Dubos, and Bookchin is that they combined humanism (so defined) with a form of naturalism that drew heavily on the ecological and evolutionary perspectives that stemmed originally from Charles Darwin. For it was Darwin who initiated the kind of ecological worldview that these social ecologists affirmed and developed, and utilised to good effect in their political critiques of industrial capitalism. For as Hans Jonas (1966) and Ernst Mayr (1988, 168–83) long ago suggested, Darwin initiated an intellectual revolution that was fundamental and far-reaching, and which had the following characteristics:

it introduced the idea that humans are not the special product of god's creation but evolved according to principles that operate throughout the natural world; it stressed the organic (not spiritual) link between humans and nature; it undermined completely—long before quantum physics (Capra), deep ecology (Naess) and eco-feminist philosophy (Plumwood)— the mechanistic world-picture, along with its dualisms, its cosmic teleology and its essentialism; it emphasised the crucial importance of openness, chance, probability, and the agency and individuality of all organisms in the evolutionary process; and, finally, it suggested ways of understanding that were both naturalistic and historical (not static and spiritual). This ecological worldview was fully embraced by all three social ecologists (Mumford, Dubos, and Bookchin) and combined with a humanistic social philosophy. Mumford described this new vision as "organic humanism," Dubos as "ecological humanism," Bookchin as "social ecology." It has affinities with the "evolutionary humanism" outlined by the biologist Julian Huxley, writing during the same period. It is however quite distinct from the ecological humanism advocated by Henryk Skolimowski (1981), who, taking his ideas from A.N. Whitehead, Teilhard de Chardin, and Martin Heidegger—all theological thinkers (he makes no mention of Darwin!)— presents us with a form of evolutionary spiritualism.

As public intellectuals, Mumford, Dubos, and Bookchin, though prolific writers, wrote in a popular style and avoided academic jargon, hoping to reach a wide audience. Although each scholar had a depth of knowledge in specific fields—Mumford wrote on architecture and urban studies, Dubos was a pioneer microbiologist, and Bookchin wrote important studies of the history of socialism and libertarian movements—all three social ecologists bewailed and critiqued the fragmentation of knowledge and the narrow specialisms that characterise contemporary intellectual life. In contrast they adopted a synthetic approach, and in their writings drew on and integrated ideas and concepts from philosophy, history, literature, anthropology, psychology, sociology, archaeology, and biology. They were radical scholars rather than academics. They thus felt that an understanding of human social life could only be attained by drawing on a multiplicity of factors—genetic, psychological, historical, environmental. They particularly aimed to bring together and integrate, in a synaptic ecological vision, the humanities (philosophy, history) and the social and biological sciences.

Neither Mumford, Dubos, nor Bookchin doubted the reality of the material world—they were realists. As Mumford expressed it: only a lunatic would fail to recognise the physical environment, and the need to breathe air, eat food, and drink water—for this constitutes the "substratum" of our daily lives. They therefore always tend to rail against idealist philosophers like Plato and Kant. But they also stressed, long before postmodernists, that our understandings of the natural world are always mediated—by our own personal experiences, and by social and cultural factors. We thus never see the world through "pristine eyes" as the anthropologist Ruth Benedict graphically put it (1934, 2). Our conceptions of nature are therefore always diverse and complex.

As pioneer ecologists, all three social ecologists offered illuminating accounts of the current ecological crisis, as well as of the social crisis. They highlighted the degradation of the natural environment under industrial capitalism—the pollution of the atmosphere and of rivers and lakes; deforestation; the limitations of industrial agriculture and the adverse effects of toxic pesticides and soil erosion; the problems of chemical additives in food; issues relating to nuclear power; and the serious decline in the quality of urban life through over-crowding, pollution, poverty and traffic congestion. Along with the economist Barbara Ward, Dubos drafted the pioneering report *Only One Earth*, which set the agenda for the United Nations Conference on the human environment (1972), and some forty years ago both Dubos and Bookchin were highlighting, with some prescience, the dangers of global warming. And long before Marxists became interested in ecological issues, Mumford, Dubos, and Bookchin were suggesting that the ecological crisis had its "roots" in an ever-expanding industrial capitalism, obsessed with economic growth and competition, a market economy that was geared to profits and power rather than human needs.

All three scholars thus came to offer radical critiques of what they describe as "industrial capitalism" or what Mumford came to portray as the "megamachine." Mumford and Dubos were essentially radical liberals, while Bookchin is a social anarchist. Nonetheless they tend to agree on the social measures that were necessary to overcome the present crisis. These include the decentralisation of the social economy, and the integration of the city and the countryside to form "bioregional" zones, thus putting an end to the "urbanisation" of the landscape; the establishment of participatory forms of democracy, involving local assemblies and direct democracy; and the scaling down of technology to a "human scale," through what later became known as "appropriate technology" (Schumacher), although Bookchin disliked the term. Along with the affirmation of craft industry, all these measures were consistent with the kind of libertarian socialism advocated by William Morris, Peter Kropotkin, and Patrick Geddes.

It is important to recognise that Mumford, Dubos, and Bookchin were not neo-Romantics; they critically engaged with, and affirmed, the Enlightenment tradition, and were not anti-technology, anti-city, or anti-science. Although critical of many aspects of modern science and technology—especially their symbiotic relationship with industrial capitalism—all three scholars affirm the crucial importance of the scientific method and of an ecologically informed technology. Unlike anarcho-primitivists, and some deep ecologists, all three scholars also positively affirmed the importance of city life-civilisation. In contrast, however, to deep ecologists and eco-philosophers, who make a "fetish" out of the "wilderness," Mumford, Dubos, and Bookchin emphasised the positive and creative aspects and the importance of humanised or cultural landscapes—which actually constitute the living environment of most humans. What they always insisted upon was the need for diversity, and thus the need to develop and conserve wilderness areas (natural landscapes), the countryside (cultured landscapes such as woods, parks, meadows, gardens, and cultivated fields), and urban settings, the town or city duly scaled to human needs and human well-being.

Given their ecological vision, Mumford, Dubos, and Bookchin always stressed that humans were an integral part of nature, and that the relationship between humans and nature should not be one of mastery of dominion but rather one that was cooperative and symbiotic—or as Bookchin expressed it, dialectical.

Providing a way to link a humanist philosophy—shorn of its arrogant anthropocentrism—with a sense of the value of nature and natural processes, the social ecology developed by Mumford, Dubos, and Bookchin continues to supply the vital philosophical underpinning to any attempt to use human ingenuity to prevent human self-destruction. From climate change to genetic modification, alternative energy sources to population growth, such a philosophy is today more relevant than ever.

References

Benedict, Ruth. 1934. *Patterns of Culture*. London: Routledge & Kegan Paul.

Bronner, Stephen Eric. 2004. *Reclaiming the Enlightenment: Toward a Politics of Radical Engagement*. New York: Columbia University Press.

Bunge, Mario. 2001. *Philosophy in Crisis: The Need for Reconstruction*. Amherst, NY: Prometheus Books.

Chomsky, Noam. 2003. *Problems of Knowledge and Freedom*. New York: The New Press.

Comaroff, John, and Jean Comaroff. 1992. *Ethnography and the Historical Imagination*. Boulder: Westview Press.

Ehrenfeld, David. 1978. *The Arrogance of Humanism*. New York: Oxford University Press.

Evans-Pritchard, E.E. 1962. *Essays in Social Anthropology*. London: Faber and Faber.

Gray, John. 2002. *Straw Dogs: Thoughts on Humans and Animals*. London: Granta Books.

Jonas, Hans. 1966. The Phenomenon of Life: Toward a Philosophical Biology. New York: Harper & Row.

Kurtz, Paul. 1983. In Defense of Secular Humanism. Amherst, NY: Prometheus Books.

Mayr, Ernst. 1988. *Toward a New Philosophy of Biology: Observations of an Evolutionist*. Cambridge, MA: Harvard University Press.

Morris, Brian. 2004. *Ecology and Anarchism: Essays and Reviews on Contemporary Thought.* Malvern Wells: Images

Skolimowski, Henryk. *Eco-Philosophy: Designing New Tactics for Living.* Boston: Marion Boyars.

Ward, Barbara, and René Dubos. 1972. *Only One Earth: The Care and Maintenance of a Small Planet.* New York: W.W. Norton.

Watson, David. 1999. Against the Megamachine. Brooklyn: Autonomedia.

7

Global Anti-Capitalism (2006)

A review of Simon Tormey's *Anti-Capitalism: A Beginner's Guide*
(Oxford: One World Publications, 2004)

A decade or so ago "globalisation" became a fashionable concept among academics. It is now a veritable "buzzword," and we are informed by the redoubtable Anthony Giddens, Blair's sociological guru (now Lord Giddens!) that there is no alternative to global capitalism. All we can do therefore is to argue about "how far, and in what ways capitalism should be governed and regulated."[1]

Capitalism has, of course, always been a global economic system, since at least the seventeenth century—a "world system" as Wallerstein described it—so it is hardly surprising that "globalisation," for many people, is simply a cover term for the latest rampant phase of capitalism. The concept, in fact, is virtually synonymous with "late capitalism," "modernisation," the "global corporate system," or what an earlier generation of socialists called "imperialism." Described as a "juggernaut," the impact of "globalisation" has indeed been profound. Essentially a political project, bolstered by a neoliberal ideology, globalisation has in recent years been implemented by the oligarchs within the IMF and World Bank, largely through what is euphemistically described as "structural adjustment programmes." This has entailed among other things: the removal of tariffs to allow transnational corporations a free hand to undermine local subsistence economies; the continued extraction of large debt repayments from Third World countries; the deregulation of international finance thus turning the global economy into a virtual "global casino"; the privatisation of land and public assets, thus enabling corporate interests to take over forests and public utilities, especially in relation to water and vital energy; the use of women and children as sweated factory labour; the replacement of subsistence crops with cash crops, thus leading to the undermining of local food security; increases in the costs of essential goods and services, such as education, transport, food, and health care; the dumping of toxic wastes in Third World countries; and last but not least, the creation of tax havens for corporate institutions and the rich.

The outcome, of course, as we all know, has been increasing poverty throughout the world, growing social and economic inequalities, and what has been described as the "dialectics of violence," reflected not only in the Iraq war, but in the disintegration of local communities, the denial of human rights, and political oppression and outright terrorism by governments. Equally, global capitalism has been the primary cause of the "ecological crisis," even threatening the basic life-support systems on which we all depend.[2]

Throughout the world there has, inevitably, been a mounting resistance to "globalisation" that has taken many different forms, although this term is somewhat misleading in describing the imperialist onslaught of global capitalism. For the ownership of capital is by no means widely dispersed; it is more centralised and concentrated than ever before in imperialist states, and although capital flows freely throughout the world seeking out profits, working people largely remain trapped within national boundaries.[3] Academic gurus like Gayatri Spivak may extol their own freedom of movement as "free spirits," but ordinary people, even asylum seekers, have no such freedom.

Even though resistance and opposition to global capitalism has a long history, the emergence of a specific global anti-capitalist movement has been dated, by many academics at least, to the protests against a meeting of the World Trade Organization that took place in Seattle in December 1999. This movement has brought together people and groups with very diverse political interests and ideologies, often conflicting—trade unions defending their jobs, a plethora of NGOs concerned with social justice, whether supporting landless peasants in Brazil or sweatshop workers in Southeast Asia, Social Democrats who resent the fact that society is being subordinated to the interests of the large corporations, Earth First! and other environmental activists, Marxists, and socialists of various persuasions, as well as many anarchist groups—primitivists, class struggle anarchists, mutualists, and the loosely affiliated group of militant anarchists known as the "black bloc," who are seen as being attached to "symbolic" violence.[4]

In the years since Seattle there have been several large demonstrations and gatherings of anti-capitalists—in Prague, Quebec City, Genoa, and in January 2001 the first World Social Forum was held in Porto Alegre, Brazil. This forum has now emerged as an intrinsic part of the global anti-capitalist movement.

During these five years numerous books have been written on the anti-capitalist movement, by academics, activists and party hacks, critiquing the corporate domination of the world's economy, and emphasising that "another world is possible"—as the slogan of the recent European Social Forum has it. The writings of Naomi Klein, Noam Chomsky, and Arundhati Roy are of particular interest in this regard—all clear and refreshingly free of academic pretensions.[5] But if you want a short, useful and thoughtful guide to the anti-capitalist movement, you can do no better than buy Simon Tormey's *Anti-Capitalism: A Beginner's Guide*. For it provides a very readable and helpful discussion of the anti-capitalist movement in all its diversity. Having taught a politics course on anti-capitalism at the University of Nottingham for some two decades, and having a mother, he tells us, who is "the original anarcho-situationist-beatnik," the book is critical, historically informed, and above all, engaging. The book however is less one of advocacy than a guide to the many issues that are relevant to any understanding of the anti-capitalist movement.

The book consists of five chapters. The opening chapter describes the nature of capitalism as an economic system—one based on private property, a competitive market, the use of capital to generate profit, and on a system of wage labour. Although he makes a distinction between the market and capitalism, Tormey tends to overemphasise the importance of the free market and competition, and thus to downplay the fact that corporate monopolies, cartels and protection rackets, and state support are intrinsic to capitalism. Indeed, Fernand Braudel made a clear distinction between the market economy and capitalism, and not only saw monopolies as being a major factor in the rise of historical capitalism, but also saw the state as a constitutive element in the functioning of the capitalist system.[6] Like Kropotkin, Braudel thus emphasised that there has always been a symbiotic relationship between the nation-state and capitalism—whether mercantile, industrial, or global.[7] The United States and the states of the European Union certainly do not practice "free trade"—this is the ideology they export to Third World countries. They themselves heavily subsidise their own transnational corporations, as well as capitalist farmers and the rich. It is equally misleading to equate capitalism with wage labour, for historically capitalism has always utilised various forms of labour—bonded labour, chattel slavery, share cropping, debt bondage, et al. The Atlantic slave trade was intrinsic to the development of capitalism, and, of course, although there has been a growing trend towards "proletarianisation," capitalists have always tended to favour "semi-proletarians," workers who lack unity and solidarity, and whose social welfare is certainly not their concern.

Tormey gives a very interesting account of the arguments that are used to justify capitalism, namely, that capitalism promotes political liberty and that its dynamism has increased the economic well-being of the majority of people—private property and economic inequalities being necessary, it seems, to make us all wealthier. Both these ideological justifications are, of course, suspect, if not vacuous.

In the second chapter Tormey discusses the background and the various factors involved in the emergence of the anti-capitalist movement. These include the following: with the emergence of global capitalism there has been a decline in the power of nation-states and thus a crisis for liberal democracy (though Tormey emphasises that the nation-state is not about to disappear); the growing disillusionment with the Soviet Union and its form of state capitalism (long critiqued, of course, by anarchists); the emergence of various new social movements since the 1960s (anti–Vietnam War, feminism, civil rights, environmentalism, nuclear disarmament); the importance of computer technology, especially the internet, which not only gives visibility to marginal groups, but also facilitates the coordination of radical activities and protests; and finally, the emergence in the 1960s of the politics of détournement—subversive activities, especially associated with the Situationists, involving guerrilla advertising and an emphasis on the aesthetic dimensions of protest. All this has given rise, Tormey suggests, to a different kind of politics—radical, diverse, extra-parliamentary, subversive, non-hierarchical. Hardly novel to generations of anarchists! Anyone who is acquainted with the history of the French Revolution and of the socialist movement (which was of course a global movement and anti-capitalist) will be aware that the politics of détournement did not begin with the Situationists; it is also important not to overemphasise the role of a small group of Situationists in the events of May 1968, or in the emergence of what Tormey calls "unofficial" politics.

When Tormey comes to discuss the nature of the anti-capitalist movement it comes as no surprise that the majority of protesters against globalisation do not in fact envisage an end to capitalism, and are not, strictly speaking, anti-capitalists. They are essentially "reformists," or Social Democrats who are committed to making capitalism work for society as a whole rather than simply in the interests of the transnational corporations. They thus seek to humanise capitalism and make it more benign, and imagine that one can have "globalisation with a human face." Chapter Three is focused on this "reformist" tendency within the anti-capitalist movement, and it is significant that Tormey devotes some thirty-four pages to the reformist anti-capitalists and only seven pages later to the anarchists. Reformism takes many different forms. Some, like Pierre Bourdieu and the Brazilian Worker's Party (who hosted the first World Social Forum in Porto Alegre) simply want to bolster the economic power of the nation-state, and thus curb the worst excesses of global capitalism;

in contrast Susan George and the French organisation "Association for the Taxation of Financial Transactions to Aid Citizens" (also one of the founders of the World Social Forum) advocate putting a tax on the movement of capital, the money collected being used to fund projects in the Third World; others still, like David Held and George Monbiot, envisage a "global social democracy" with global institutions and a global state to ensure that global capitalism serves the interests of the majority of people. Monbiot still suffers from the liberal illusion that capitalism in not intrinsically exploitative, and that the democratic state essentially serves the public interest. Tormey suggests that the global Social Democrats are akin to the Jacobins of the French Revolution and that their radical vision is essentially utopian.

In Chapter Four, entitled "Renegades, Radicals and Revolutionaries," Tormey discusses the radical wing of the anti-capitalist movement. It consists of five broad political tendencies, as outlined by Tormey. These are the Marxists, with their emphasis on the primacy of productive relations, and the necessity of a party organisation for defeating capitalism (the Socialist Worker's Party is prototypical of this tendency); the Autonomists, those Marxists who stress the primacy of political struggles and have abandoned the idea of a vanguard party in favour of worker's councils (Anton Pannekoek's "council communism," the Situationists, and the writings of such intellectuals as Antonio Negri are seen as reflecting this tendency)[8]; the anarchists, whom Tormey describes as consisting of a "staggeringly diverse range of political currents and groups"; the radical Greens under whose rubric Tormey mentions Bookchin and Zerzan; and finally, the Ejército Zapatista de Liberación Nacional (EZLN) or Zapatistas, the army of national liberation that took over the Chiapas region of Mexico in 1994, allowing the establishment of autonomous peasant communities in the region. Tormey suggests that the Zapatista movement is an entirely novel phenomenon, and like the radical Greens, is "beyond" ideology or "non-ideological." Peasant movements and peasant resistance to an encroaching capitalism are hardly new phenomena, and the notion that some movement or organisation is "non-ideological" is itself profoundly ideological. The ideology of the Zapatistas is clearly evident in the writings of that shadowy figure Subcomandante Marcos, who seems to have become the sole "spokesperson" for the indigenous peasant communities of the Chiapas. His letters and declarations to the Mexican people and the world, emphasise the rights of the indigenous people of Mexico to land and liberty, articulate as central demands democracy and social justice, and emphasise the need to restore the national sovereignty of the Mexican state.[9]

Unfortunately with regard to anarchism, Tormey evinces some rather quaint ideas, and two are noteworthy: firstly, that Karl Marx was a true anarchist, unlike Proudhon and Bakunin, whom Marx, we are told, continually berated for their political limitations; secondly, that many anarchists are supporters of capitalism.

It has often been suggested that Marx and anarchists like Bakunin were in fundamental agreement as to the aims of the revolutionary movement, in that they both envisaged a future society that would be socialist and stateless. Maximilien Rubel, in fact, even suggested that Marx was not an authoritarian socialist but rather the first "to develop a theory of anarchism." But as many people have argued, the aims of Marx and Bakunin were quite dissimilar, in that the future envisaged by Marx looked to a high degree of industrial technology, with a corresponding degree of centralised institutions through which the social and economic life would be "managed." The state, for Marx and Engels, would "administer" society. As they put it in their address to the central committee of the Communist League (1850): the workers must strive to create a German republic, and within this republic strive "for the most decisive centralization of power in the hands of the state authority." No wonder Bakunin, with some prescience, saw this as inevitably leading to the emergence of a highly despotic government. Marx was certainly no anarchist in his politics, whatever his vision of a future communist society.[10]

What about anarchists who supposedly support capitalism? Tormey mentions individualist anarchists like Tucker, Spooner, and Warren, and anarcho-capitalists like Ayn Rand and David Friedman. Although the nineteenth-century individualist anarchists or mutualists advocated private property, petty commodity production, and a market economy, it is questionable, given their emphasis on liberty, whether they would have supported the kind of global capitalism that is now rampant throughout the world. It is evident from their writings that they rejected both capitalism and communism—as did Proudhon. As for Ayn Rand being an anarchist, the suggestion is quite bizarre. Not only Margaret Thatcher's guru and a strident advocate of laissez-faire capitalism, Rand was also an advocate of the "minimal" but highly repressive state, necessary to support private property rights.[11] Anarcho-capitalists like Friedman and Murray Rothbard simply replaced the state with private security firms, and can hardly be described as anarchists as the term is normally understood.

In the final chapter Tormey offers some interesting reflections on the future of the global anti-capitalist movement. He seems to have a strong predilection to the idea of "novelty," although of course all social phenomena exhibit both continuity and change. He writes as if social networks never existed until the computer was invented or was theorised by fashionable academic icons like Deleuze. But he seems to acknowledge that demonstrations against the Vietnam War or the large demonstrations focused around environmental issues were not simply an expression of "identity" politics but were in essence both global and anti-capitalist, bringing together people of diverse political persuasions. Drawing a stark contrast between democratic, official, party politics with its emphasis on the "capturing of power"—vertical and ideological politics—and "network" or "rhizomatic" politics with the emphasis on social networks—horizontal, spontaneous, disorganised, and transient—Tormey seems to bypass entirely the strategy that has long been that of the anarchists, namely, the support and creation of voluntary organisations. The very term "social" implies enduring social relationships, and without some form of social life—village communities, local assemblies, workers' councils, producers' cooperatives, housing or neighbourhood associations, affinity groups, anarchist federations—one could not even obtain one's daily bread, let alone decide how to cooperatively produce it. The Zapatistas certainly do not live in a world of computers and disembodied spontaneous "networks." But overall Tormey's book is an excellent guide to the global anti-capitalist movement and has a useful glossary and a chronology of the events and initiatives relating to the movement since 1998.

Notes

1   Anthony Giddens, *The Third Way* (Cambridge: Polity Press 1998), 43.
2   Wayne Ellwood, *The No-Nonsense Guide to Globalization* (London: Verso, 2001), 107.
3   *The Economics and Politics of the World Social Forum* (Mumbai: Research Unit for Political Economy, 2003), 5.
4   See "Perspectives on Anti-Capitalism" in *Organise!* 56 (2001); Emma Bircham and John Charlton, *Anti-Capitalism* (London: Bookmarks, 2001), 271–83.
5   Naomi Klein, *No Logo* (London: Flamingo, 2000); Noam Chomsky, *Profit over People* (New York: Seven Stories Press, 1999); Arundhati Roy, *The Ordinary Person's Guide to Empire*, (London: Flamingo, 2004).
6   See Immanuel Wallerstein, *Unthinking Social Science* (Cambridge: Polity Press 1991), 205; and my study *Kropotkin: The Politics of Community* (New York: Humanities Books 2004), 202–3.
7   Immanuel Wallerstein, *Historical Capitalism* (London: Verso, 1983), 27.
8   See Michael Hardt and Antonio Negri, *Empire* (Cambridge, MA: Harvard University Press, 2000).
9   Juana Ponce De Leon, *Our Word Is Our Weapon* (New York: Seven Stories Press 2001).
10   On these issues see Karl Marx, *The Revolutions of 1848* (London: Penguin, 1973), 328; Brian Morris, *Bakunin: The Philosophy of Freedom* (Montreal: Black Rose Books, 1993), 125–35; Paul McLaughlin, *Mikhail Bakunin: The Philosophical Basis of His Anarchism* (New York: Algora, 2002), 76–78.
11   On Ayn Rand see my study *Ecology and Anarchism* (Malvern Wells: Images, 1996), 183–92.

8

People without Government (2007)

Two Images of Humans

Western social science and eco-philosophy are perennially torn between two contradictory images of the human species. One, associated with Thomas Hobbes (1651), sees human social life as a "war against all," and human nature as essentially possessive, individualistic, egotistic, and aggressive. It is a basic tenet of the "possessive individualism" of liberal political theory (MacPherson 1962). The other, associated with Rousseau, depicts human nature in terms of the "noble savage"—of the human species as good, rational, and angelic, requiring only a good and rational society in order to develop their essential nature (Lukes 1967, 144–45). Both these images are still current and have their contemporary exemplars: the Hobbesian image is evident in the accounts of many psychoanalysts and sociobiologists, while Nance's study of *The Gentle Tasaday* (1975) suggests a highly romantic picture of a peaceful, nonviolent society. In the writing of many eco-feminists and Afro-centric scholars, a past "golden age" is portrayed—in which peaceful social relations, gender equality, and a harmony with nature pertained, before the rise respectively of bronze age culture and colonialism (Eisler 1987, Diop 1989). Both these images share, of course, a similar theoretical paradigm which sees human relations as solely "determined by some natural state of human beings" (Robarchek 1989, 31). The contributors of the volume *Societies at Peace* (Howell and Willis 1989) all eschew, along with Robarchek, this biological determinism, and emphasise an approach that dispenses with "universalistic definitions," suggesting that human behaviour is never culturally neutral, but always embedded in a shared set of meanings. Yet they argue strongly that "sociability" is an inherent capacity of the human species, and all the essays tend towards the tradition of Rousseau. But countering biological and deterministic approaches to culture should not lead us to endorse an equally one-sided cultural (or linguistic) determinism that completely oblates biology.

What Is Politics?

Anthropologists' past contribution to political science, it has been said, focuses specifically on two important fields. One is in outlining the politics of societies without centralised government, and studies by Malinowski on the Trobriand Islands and Evans-Pritchard on the Nuer have become classics. The other is in the analysis of micro-politics, particularly of political leadership, village politics, and the relationship between politics and symbolism (Bailey 1969, Cohen 1974). Here I shall focus on the first of these contributions, *People without Government*. This is the title of a very useful book by Harold Barclay (1982), who writes from an anarchist perspective. The book is, indeed, subtitled *An Anthropology of Anarchism*.

Order and power are intrinsic to social life. A human society has, by definition, both order and structure, and operates with regularised and relatively fixed modes of behaviour. Humans without society are not human, for society is basic to the human condition, as Marx long ago insisted (see Carrithers 1992). So is power. Power is a relationship, and implies the ability "to get others to do what you want them to do." Power may mean influence—convincing others by monetary rewards, by logical argument, or by the prestige of one's status. Or it may mean coercion—the implied or overt threat of injury. But power is intrinsic to any social group. The question for anarchists, therefore, is not whether there should be order or structure, but, rather, what kind of social order there should be, and what its sources ought to be. Equally, anarchists are not utopians who wish to abolish power, for they recognise that power is intrinsic to the human condition. As Bakunin expressed it: "All men possess a natural instinct for power which has its origin in the basic law of life enjoining every individual to wage a ceaseless struggle in order to insure his existence or to assert his rights" (Maximoff 1953, 248).

What anarchists strive for is not the abolition of power but its diffusion, its balance, so that ideally it is equally distributed (Barclay 1982, 16–18). The notion that anarchists endorse unlimited freedom, as Andrew Heywood suggests (1994, 198), is a serious misunderstanding of anarchism. Anarchists affirm with J.S. Mill, who, in *On Liberty*, argued that the only justification for interfering with the liberty of some person is to prevent physical harm being done to oneself or others (Mill 1972, 78). Anarchism does not imply licence; rather it repudiates coercive power.

Authority, as Weber long ago explored (1947), is power that is considered legitimate by members of a community. But, as Barclay stresses, such legitimacy may be more in terms of "tacit acquiescence" rather than in the unconditional acceptance of power, and, citing Morton Fried, he notes that legitimacy is the means by which ideology is harnessed to support power structures.

The function of legitimacy is "to explain and justify the existence of concentrated social power wielded by a portion of the community and to offer similar support to specific social orders, that is, specific ways of apportioning and directing the flow of social power" (Fried 1967, 26). All human societies, therefore, have political systems, but not all have government, for the latter is but one form of political organisation.

In the preface to the classic survey *African Political Systems* (1940), A.R. Radcliffe-Brown defines political organisation in terms of "the maintenance or establishment of social order, within a territorial framework, by the organised exercise of coercive authority through the use, or the possibility of use, of physical force" (xiv). He went on to suggest that the political organisation of a society "is that aspect of the total organisation which is concerned with the control and regulation of the use of force" (xxiii).

Such a definition, which is clearly derived from Weber in its dual stress on territory and coercive force, essentially refers to government, and is thus too limiting as a definition of politics. Weber had defined "power" (*macht*) as the "probability that one actor within a social relation will be in a position to carry out his own will despite resistance," and defined a group as "political": "if and in so far as the enforcement of its order is carried out continuously within a given *territorial* area by the application and threat of physical force" (1947, 152–54).

Fortes and Evans-Pritchard, in their introduction to *African Political Systems*, found such definitions of politics too restrictive, and noted that ethnographers who, like themselves, studied such societies as the Nuer and Tallensi—societies which lacked centralised authority—were forced to consider "what, in the absence of explicit forms of government, could be held to constitute the political structure of the people" (1940, 6). In the study a simple division is made, following a tradition established by Maine and Morgan, between two main categories of political system: those societies having centralised systems of authority—that is, having a government or state (societies such as the Bemba or Zulu), and those societies which lack centralised authority, such as hunter-gatherers and the aforementioned Tallensi and Nuer.

Although acknowledging that there is an intrinsic connection between people's culture and their social organisation, Fortes and Evans-Pritchard emphasise that these two components of social life must neither be confused nor conflated. They note that culture and type of political system vary independently of one another, and that there is no simple relation between modes of subsistence and a society's political structure. But they acknowledge that, in a general sense, modes of livelihood determine the dominant values of a people and strongly influence their social organisations including their political systems. They suggest that wide divergences in culture and economic pursuit may be incompatible with what they describe as a "segmentary political system," characteristic of the Nuer, Tallensi, and Logoli. In the latter system there is no administrative organisation or government, and the "local community," not the state, is the key "territorial" unit. Membership of the local community, they suggest, is acquired as a rule through genealogical ties, whether real or fictitious, and they write: "The lineage principle takes the place of political allegiance, and the interrelations of territorial segments are directly coordinated with the interrelations lineage segments" (Fortes and Evans-Pritchard 1940, 11). The political structure of these societies thus consists of an "equilibrium between a number of segments, spatially juxtaposed and structurally equivalent, which are defined in local and lineage, and not in administrative, terms ... [it] is a balance of opposed local loyalties and of divergent 'lineage and ritual ties.'" (13).

Middleton and Tait's important study *Tribes without Rulers* (1958) is devoted to outlining the political structure of six African societies where political authority is focused around local groups that are united by unilineal descent. These societies—Tiv, Mandari, Dinka, Bwamba, Kankomba, and Lugbara—are thus seen as having an uncentralised political structure based on a "segmentary lineage system." The authors recognise that in other African communities political authority may be invested in other institutions, such as local village chiefs, age-set systems, village councils or associations, or ritual fraternities, and that these are often found in conjunction with lineage structures. They recognise, too, the diversity of these six societies, with respect to the existence of chiefs, and whether or not the descent groups are dispersed. But in all these societies, they suggest, the "segmentary principle" is operative, namely that political relations between territorial groups are conceived in terms of descent, whether of a lineage or clan system.

There has been much critical discussion of this mode of analysis, and of "descent theory" more generally. The assumed and simple correlation between local groups (land), membership of descent groups (kinship) and political affiliation has been critically questioned, especially in relation to the Nuer. The whole notion of a "gentile organisation," a polity based on exogamous descent groups, has been described as more "mythical" than real by one scholar (Beidelman 1971; Kuper 1988, 190–209; but see Kelly 1985 on Nuer expansion).

The simple equation of politics with hierarchy and coercive power was also challenged by Pierre Clastres in his classic study *Society against the State* (1977). Like Barclay, Clastres belongs to a long anarchist tradition that goes back to the end of the eighteenth century. The study is focused—according to the dust jacket—on the "leader as servant and the human uses of power among the Indians of the Americas." The book is appropriately entitled *Society against the State*, for, like Tom Paine and the early anarchists, Clastres makes a clear and unambiguous distinction between society and the state, and suggests that the essence of anarchic societies, whether hunter-gatherers or early Neolithic peoples, is that effective means are institutionalised to prevent power being separated from social life.

The classical definition of political power in the Western intellectual tradition, evident in the writings of Nietzsche and Weber, as well as by anthropologists—quoted above—put a fundamental emphasis on control and domination. Power is always manifested within "a relationship that ultimately comes down to coercion . . . the truth and reality of power consists of violence" (1977, 4). The Western model of political power, which stems from the beginning of Western civilisation, tends to see power in terms of "hierarchized and authoritarian relations of command and obedience" (9). Such a viewpoint, Clastres feels is "ethnocentric," and it immediately leads to puzzlement by ethnologists when they confront societies without a state, or without any centralised agencies. Such societies are conceptualised as missing something, as incomplete, as lacking . . . a state. In social contexts where there is neither coercion or violence, is it then possible to speak of political power? Scholars have thus been led to describe "power" in the Trobriand Islanders or such societies as Nilotic people of the Sudan as being "embryonic" or "nascent" or as "undeveloped." History is then seen as a "one-way street," with Western culture seen as the "image" of what "societies without power will eventually become." But Clastres contends that there are no human societies without power. What we have is not a division between societies with power and societies without power (stateless societies)—for "political power is universal, immanent to social reality" (14)—but rather a situation in which power manifests itself in two modes—coercive and non-coercive. Political power is thus inherent in social life; coercive power is only a particular type of power.

Clastres notes how the first European explorers to South America were bemused and bewildered in describing the political life of the Tupinamba Indians—"people without god, law or king"—but felt at home in the hierarchic states of the Aztecs and Incas—with their coercive and hierarchic political systems. For Clastres, then, political power as coercion or violence is the stamp of *historical* societies, and it is the political domain itself which constitutes the "first motor" of social change.

In examining the philosophy of the Indian chieftainship, Clastres argues that the chiefs lacked any real authority, and that most Indian communities of South America, apart from the Incas, were distinguished by "their sense of democracy and force for equality" (20). Reviewing the ethnographic literature, Clastres suggests that four traits distinguished the chief among the forest tribes of South America. Firstly, the chief was a peacemaker, responsible for maintaining peace and harmony within the group, though lacking coercive power. His function was that of "pacification," and only in exceptional circumstances, when the community faced external threat, was the model of coercive power adopted. Secondly, the chief must be generous with his possessions; as Clastres quotes from Francis Huxley's study of the Urubu, you can always recognise a chief by the fact that he has the fewest possessions and wears the shabbiest ornaments (22). Thirdly, a talent for oratory, Clastres suggests, is both a condition and an instrument of political power, such oratory being focused upon the fundamental need of honesty, peace and harmony within a community. Fourthly, in most South American societies, polygamous marriage is closely associated with chiefly power, and it is usually the chiefs prerogative, although successful hunters may also have polygamous marriages. As polygamy is found among both the nomadic Guayaki and Siriono, hunter-gatherers in which the band rarely numbers more than thirty persons, and among sedentary farmers like the Guarani and Tupinamba, whose villages often contain several hundred people, polygamy is not an institution that is linked to demography, but is rather linked to the political institution of power.

All these traits are fundamental expressions of what constitutes the basic fabric of archaic society, namely that of exchange. Coercive power, Clastres suggests, is a negation of this reciprocity. Accepting Murdock's contention that the atavism and aggressiveness of tribal communities has been grossly exaggerated, Clastres highlights the importance of marriage alliances, especially cross-cousin marriages, in establishing "multicommunity" structures. He refers to these as "polydemic structures" (53). He also emphasises that among the Guayaki (Aché) foragers there is a fundamental opposition between men and women, whose economic activities form two separate but complementary domains, the men hunting and the women gathering. Two "styles" of existence are thus seen to emerge, focused on the cultural opposition between the bow (for hunting) and the basket (for carrying), which evokes specific reciprocal prohibitions. Importantly, for the Guayaki hunter, there is a basic taboo that categorically forbids him from partaking of the meat from his own kill. This taboo, Clastres suggests, is the "founding act" of an exchange of food which constitutes the basis of Guayaki society.

Clastres emphasises the fact that a subsistence economy did not imply an endless struggle against starvation but rather an abundance and variety of things to eat, and that, as with the Kalahari hunter-gatherers, only three or four hours were spent each day in basic subsistence tasks—as work. These communities were essentially egalitarian, and people had a high degree of control over their own lives and their work activities. He argues that the decisive break between "archaic" and historical societies was not the neolithic revolution, and the advent of agriculture, but rather stems from a "political revolution," the emergence of the state. The intensification of agriculture implies the imposition, on a community, of external violence. But such a state apparatus is not derived, Clastres argues, from the institution of chieftainship, for in archaic societies the chief "has no authority at his disposal, no power of coercion, no means of giving an order" (174). Chieftainship thus does not involve the functions of authority. Where then does political power come from? Clastres tentatively suggests that the origins of the state may derive from religious prophets, and concludes by noting that while the history of historical society may be the history of class struggle, for people without history it is "the history of their struggle against the state" (186).

The key point of Clastres's analysis, later confirmed by John Gledhill (1994, 13–15), is that it provides a critique of Western political theory which tends to identify political power with violence and coercion, as well as highlighting an important lesson to be derived from anthropology, namely that it is possible for societies to be organised without any division between rulers and the ruled, between oppressors and the oppressed. It also suggests that we look at history not in terms of typologies, but rather as a historical process where, within specific regions, societies with states have co-existed with "stateless" populations which have endeavoured to maintain their own autonomy and to resist the centralising intrusions and exploitation inherent in the state (Gledhill 1994, 15). It is also worth noting that anarchists have always made a distinction, long before Deleuze, between organisation and order imposed from above.

Societies without Government

An important tradition within anthropology has been to interpret the political systems of pre-capitalist societies in terms of typologies that are essentially taxonomic and descriptive. Following the earlier neo-evolutionary approach to politics, associated with Elman Service (1962) and Morton Fried (1967), Ted Lewellen (1992) has suggested four types of political systems, based on their mode of political integration.

The BAND type of political organisation is characteristic of hunter-gathering societies like the !Kung of the Kalahari, the Inuit, and the Mbuti of Zaire, as well as of all prehistoric foragers.

TRIBES: Although Lewellen notes the problematic nature of the concept of "tribe," he advocates the use of the term on both logical and empirical grounds. In evolutionary terms there must be some political term that is midway between the "band" level of political organisation associated with hunter-gatherers, and centralised political systems. Cross-cultural systems also do reveal certain features which "tribal" societies do have in common, although they also show wide variations with respect to the existence of age-sets, pan-tribal sodalities, and ritual associations. Lewellen outlines the political in three tribal contexts, that of the Kpelle, Yanomami, and Nuer, and also considers the Iroquois to exemplify this type of political system.

CHIEFDOMS transcend the tribal level in having some form of centralised system and a higher population density made possible by more efficient productivity. There may be a ranked political system, but no real class differentiation. Lewellen describes the Kwakiutl and pre-colonial Hawaii as being typical chiefdoms.

Finally, there is the STATE level of political integration, which implies specialised institutions and centralised authority in order to maintain by coercive force differential access to resources. The key feature of the state is its permanence. Lewellen gives a descriptive outline of the pre-colonial Inca and Zulu states.

Here I shall examine three contexts: foragers, small-scale horticulturists who live in villages, and chiefdoms.

In an important review of the literature, Marvin Harris (1993) emphasises the salience of bio-sexual differences in the understanding of gender hierarchy in human societies. The basic differences between men and women, in terms of stature, musculature, and reproductive physiology—only women become pregnant and lactate—provides, he suggests, a "starting point" in attempting to understand gender. Cultural determinism therefore does not counsel us to ignore biology, nor does the emphasis on biological difference imply a simple biological determinism such that "anatomy is destiny."

Such biological differences, Harris suggests, are clearly related to one of the most ubiquitous features of early human societies—both contemporary hunter-gatherers and prehistoric foragers—namely the division of labour by sex. With few exceptions, such as that of the Agta of Luzon—where women hunt wild pigs and deer with knives and bows and arrows (see Dahlberg 1981)—among hunter-gatherer societies men are primary hunters of large game. They thus become specialists in the making of hunting weapons, such as bows and arrows, spears, harpoons, boomerangs, and clubs—weapons that could also be used to injure or kill other humans. But the association of men with hunting, and with the control of weapons, did not necessarily entail gender hierarchy. There is plenty of evidence to suggest that among many foragers (and some subsistence cultivators) the sexual division of labour is complementary, and gender relations are essentially egalitarian, as Clastres implied. Also, in early human communities, scavenging and group hunting by all members of the community was probably widespread (Ehrenreich 1997).

Harris cites the studies of Eleanor Leacock among the Montagnais-Naskapi foragers of Labrador, Turnbull's studies of the Mbuti of Zaire, and Marjorie Shostak's biography of Nisa, a !Kung woman, to indicate that women in foraging societies have a high degree of autonomy, and that egalitarian relations between the sexes is the norm (Leacock 1983, Turnbull 1982, Shostak 1981). But Harris deems that gender roles in foraging societies aren't completely complementary and egalitarian, for in their role as healers, and in the realm of public decision-making, men often tend to have a "significant edge" over women in almost all foraging contexts (1993, 59). Although organised violence is not found among the !Kung of the Kalahari, Harris argues that they are by no means the "peaceful paragons" as depicted by Elizabeth Marshall Thomas in her book *The Harmless People* (1958). Violent arguments frequently occur, and homicide is not unknown. Significantly, Richard Lee found that in thirty-four cases of inter-personal conflict over a five-year period—half of which involved domestic disputed between spouses—it was the man who initiated the attack in the majority of cases, and of the twenty-five cases of homicide, though the victims were mainly men, all the killers were also men (Lee 1979, 453). Citing one comparative study (Hayden et al 1986), Harris suggests that where conditions entail the development of feuding among hunter-gatherers, then this correlates with an increased emphasis on male dominance—for then a warrior ethic and male aggressiveness is given cultural prominence.

Warfare is organised conflict involving teams of armed combatants; among the !Kung however, such warfare did not exist, and there was a virtual absence even of raiding. This is consonant with a situation where gender equality is the norm. Yet, as Harris suggests, many band-level societies engage in inter-group warfare to varying degrees, and thus possess well developed forms of gender hierarchy. He also cites the ethnographic accounts of the Australian Aborigines, although also noting that in these societies women had a considerable degree of independence.

Besides an ethos of sharing, a complementarity of gender rites, and a general level of gender equality among foraging societies (see Woodburn 1982, Kent 1993), there is also an important emphasis on consensus. This is clearly brought out in George Silberbauer's seminal essay on the G/wi (1982).

The G/wi speaking Bushmen of the Central Kalahari, Botswana, were studied by Silberbauer between 1958 and 1966, when they were still largely autonomous hunters and gatherers. Since then, the region has been increasingly penetrated by Tswana and Kgalagadi pastoralists.

The social and political community of the G/wi is the band, which is conceptualised in terms of a group of people living in a specific territory and controlling the use of its resources. Membership of the band is primarily through kinship and marriage, but membership is open and not exclusive, so non-G/wi can become members. Within the band there is movement and flux, and a continuing pattern of separation and integration between the various householders that constitute it. This enables the local group to successfully exploit environmental resources. To do this, Silberbauer suggests, political processes must be "integrative without weakening inter-household dependence which would cripple the autonomy" of the household—for people's survival depends on this autonomy. Kinship, which has universalistic properties, is important in ordering relationships within the band.

Decisions affecting the band as a whole are arrived at through discussions, involving all adult members. Such discussions tend to be informal, and seldom take the form of set-piece public debates. Disputes and arguments are addressed in public, but these are done indirectly, as direct confrontation between opposing individuals is seen as a breach of etiquette. During the summer and autumn, joint camps are formed, but these are unstable groupings, and their composition is always based on a preference for one another's company. These groupings—or "cliques" as Silberbauer calls them—form an ephemeral segmentation of the band.

Leadership of the band is evident at all phases of decision making, which is initiated by someone identifying or communicating a problem that needs a resolution. Leadership is apparent to the degree that someone's suggestion or opinion attracts public support, and it shifts according to context or relevant expertise. Public decisions cover a wide field, ranging from domestic disputes to the location of the next camping site. Decisions are essentially arrived at by consensus, but this by no means entails a unanimity of opinion or decision. It rather implies a situation where there is no significant opposition to a proposal. All members of the band have the opportunity to participate in the decision. As consensus implies an element of consent, it negates the notion of coercion—and the general openness of the band as a social unit prevents coercive factions from emerging.

Silberbauer thus concludes that the style of band politics is facilitative rather than coercive, and leadership is authoritative rather than authoritarian, an individual striving for the cooperation of others in the activities they may wish to undertake. He distinguishes such consensus politics from a democracy—which involves equal access to positions of legitimate authority, and is essentially an organisational framework for the making and execution of decisions. Silberbauer suggests that the common definition of political action is terms of coercive power or physical force, suggested by Max Weber (1947, 154) and Radcliffe-Brown (1940, xxiii), noted above, is too narrow and selective, and is inappropriate in the context of consensus politics. It leads, he suggests, "to the paradox that, as there is no locus of power, such a polity has no authority. This is, of course, nonsense for it is the very fact of consensus which lends authority to the decision" (1982, 33).

A second context discussed by Harris is that of village organised societies, where subsistence is derived in part from rudimentary forms of agriculture, and where armed raiding is almost endemic. The two classic contexts are the Yanomami of Venezuela—the subject of important studies by Chagnon (1968) and Lizot (1985), and the village communities of the New Guinea highlands. The Yanomami, described as the "Fierce People" by Chagnon, train boys from an early age to become warriors, to be courageous, cruel, and vengeful. Young boys learn their aggression and cruelty by practising on animals. Armed raids are undertaken at dawn on rival villages, and women taken as captives. Successful men are polygamous, and there is a pervasive pattern of ill-treatment towards women, who are beaten and harassed. About a third of the deaths in some Yanomami villages result from armed combat, and the overall homicide rate is high—five times greater that of the !Kung (Knauft 1987, 464).

The abuse and mistreatment of women is equally evident among many New Guinea communities, who, according to Harris, are the "world's most ardent male chauvinists" (1993, 65). The central institution of these societies, the Nama, which is a male initiation cult, essentially trains men to be fierce warriors, and to subordinate women. Among the Sambia, as described by Gilbert Herdt (1987), there is a rigid segregation of the sexes, the men being engaged in fighting and hunting, the women tending to the pigs, and doing what Herdt describes as the "routine" cultivating of the gardens. Men avoid all contact with children, and fear intimacy with women, their main activities being focused around the secret male clubhouse. Through complex initiations boys become members of what Herdt calls a "clan-based warriorhood," centred on a local hamlet. Through ritual fellatio, semen is passed from men to boys, and the loss of semen through heterosexual activities is feared—as contact with women is believed to be "polluting." Sexual antagonism is therefore characteristic of Sambia relationships, and constitutes for them a psychological reality. The coordinating institution of this patrilineal society is the men's secret society; it is a dominating force in Sambia social life, and an instrument of political and ideological control of men over women.

But not all village based communities who practise horticulture—with hunting as an important subsidiary activity—are characterised by male dominance and an ethic of violence. By way of contrast, therefore, it is worth briefly recounting two interesting essays from the collection on *Societies at Peace* by Joanna Overing and Clayton Robarchek respectively.

While Marvin Harris tends to conceptualise village based communities in the tropics as essentially focused around a warrior ethic, with coercive relationships and male dominance as correlates, Pierre Clastres's account of South American forest tribes emphasises their egalitarian ethos, and their aversion to coercion and hierarchy. Joanna Overing (1989) brings these two contrasting perspectives together in her account of the Shavante and Piaroa "Styles of Manhood."

The Shavante of Central Brazil, studied by Maybury-Lewis (1971) have a gathering economy, supplemented by both hunting and horticulture. But hunting is more than simply an economic activity, though both men and women have a passion for meat, for hunting is intrinsically linked with male sexuality, providing the hunter with "a public stage" for a stylised display of virility. Masculinity is thus defined in terms of self-assertiveness, violence and a belligerent temper—such belligerence being instilled in boys from an early age. Gender antagonism or "sexual bellicosity" is thus intrinsic to the Shavante definition of manhood, and ritual violence against women is elaborated. Men have political supremacy, and violence occurs both within the community, and in hostilities with outsiders. According to Maybury-Lewis, much of Shavante life is a function of politics, and such politics is based on competition between groups of males (1971, 104).

Overing notes that this description of the Shavante is in accordance with Collier and Rosaldo's (1981) depiction of the culture of a "bride service society," where hunting, killing and male sexuality are ideologically linked—a depiction, she feels, which is based on a rather selective examination of the ethnographic material.

The Piaroa style of "manhood," Overing suggests, stands in "extreme contrast" to that of the Shavante living in Southern Venezuela. The Piaroa, like the Shavante, combine gathering with hunting and garden cultivation—as well as fishing. They are, relatively speaking, highly egalitarian, although each territory has a politico-religious leader (Ruwang), though his authority is limited. Neither the community, as a collective, nor any individual, "owns" land, and all products of the forest are shared equally among members of the household. Piaroa social life, according to Overing, is very unformalised, and a great emphasis is put on personal autonomy. They see great virtue in living peacefully, and in being "tranquil," and their social life is almost free of all forms of physical violence. Coercion has no place in their social life, and any expression of violence is focused on outsiders. Gender relations are neither hierarchic nor antagonistic, and the ideal of social maturity is the same for both men and women—one of "controlled tranquillity" (87). The portrait of Piaroa society thus accords with that suggested by Clastres.

The Semai people of Malaysia were the subject of an important early study by Dentan (1968)—who significantly described them as a "non-violent" people. In recent years they have been portrayed, Robarchek (1989) suggests, in terms of both the images that we earlier described—as both the quintessential "noble savage," and as bloodthirsty killers. Robarchek, in his ethnographic account of these people, whose social life is seen as "relatively free of violence," steers between these two extremes, and sees the Semai as an example of a "peaceful society"—along with the Mbuti of the Ituri forest, the Kalahari Bushmen, the Tahitians, the Inuit and the Haluk (Turnbull 1961, Thomas 1958, Levy 1973, Briggs 1970, Spiro 1952). But the emphasis on nonviolence does not necessarily imply a lack of egoism or individualism, and Robarchek suggests that among the Semai there is a psycho-cultural emphasis on individualism and autonomy, as well as on nonviolence, nurturance and dependency—a theme I explored in my study of another Asian forest community, the Hill Pandaram (Morris 1982).

The themes of danger and dependency, according to Robarchek, are ubiquitous in Semai's social life. Danger is felt to be omnipresent—from the natural world, from spirits, and from outsiders. However, Robarchek does not explore the socio-historical context of the Semai; encapsulated as they are within a wider economic system, they are people who have, through the centuries, been harassed and exploited by outsiders. Dependency has equal emphasis, and there are important moral imperatives to share food, and to avoid conflict and violence. Paramount emphasis is thus given to the values of nurturance, generosity, and group belonging. The protection and nurturance by the kin community is described as "the only refuge" in a hostile world—although the dangers are expressed by Robarchek in terms of a cultural image rather than as stemming from a political reality.

But this emphasis on sharing, dependency and nonviolence co-exists with an equally important emphasis on individual autonomy. A sense of individuality, of personal autonomy, and of freedom from inter-personal constraints, is stressed from the earliest years of childhood—and at extremes this may entail for the Semai emotional isolation, fragility in marriage ties, and a lack of empathy towards others.

Other Asian forest people have been described as "peaceful societies," and exemplify a similar cultural pattern to that of the Semai. In her account of the Chewong, for example, Signe Howell (1989) suggests that for these people, as the title of her article suggests, "To be angry is not to be human, but to be fearful is." On the basis of the ethnographic data, she questions whether "aggression" is an intrinsic part of human nature. Gibson, likewise, in his discussion of the Buid of the Philippines—also shifting cultivators like the Semai and Chewong—suggest that these people are a society "at peace," for they place a high moral order on "tranquillity," and a corresponding low value on "aggression." But Gibson sees these moral attitudes as the product of historical processes in which the Buid were consistently the victims of outside forces. Their culture cannot therefore be seen simply as an effect of innate psycho-biological capacities, nor in terms of their adaptation to the forest environment (1989, 76).

Among hunter-gatherers, and such village-based, small-scale horticulturists as the Yanomami, Semai, and Sambia, there is close correlation between the degree of internal warfare—armed raids—and the degree to which gender hierarchies develop, the degree that is, of male domination over women. But this correlation does not hold, Harris suggests, when we move to societies with a more complex political system, those constituting "chiefdoms." Such chiefdoms typically engage in warfare with distant enemies, and this, he writes, "enhances rather than worsens the status of women since it results in avunculocal or matrilocal domestic organisations" (1993, 66).

In more complex, multi-village chiefdoms, where men undertake long sojourns for the purposes of hunting, trade, or warfare, matrilocality tends to prevail. In this context women assume control over the entire domestic spheres of life. External warfare is therefore associated, Harris suggests, with matrilineal kinship and a high degree of gender equality.

The classic example of this association of external warfare and gender equality—Harris puts an emphasis on warfare rather than on hunting or external trade—is that of the Iroquois. These matrilocal, matrilineal people resided in communal long houses whose activities were directed by senior women. The in-marrying husband had little control over domestic affairs, agriculture being largely in the hands of the women. The political system of the Iroquois consisted of a council of elders, of elected male chiefs from different villages. Senior women of the long houses nominated the members of this council, but they did not serve on the council. However, they could prevent the seating of any man that they opposed, and by controlling the domestic economy had a great deal of influence over the council's decisions. In the public domain they thus possessed by indirect means almost as much influence as men (Brown 1975). However, this situation did not entail a matriarchal situation, Harris contends, for the women did not humiliate, exploit, or harass their men. This, however, had little to do with their feminine nature: there is plenty of evidence of women's involvement elsewhere in armed combat, and of them being enthusiastic supporters of war and torture. "It was lack of power and not lack of masculinity," Harris writes, that prevented women in pre-industrial societies setting up matriarchal systems (1993, 69).

In his lively, readable text *Cannibals and Kings* (1977, 92–93), Harris suggests that matrilineal forms of organisation were a short-lived phase in the development of primitive states. He writes: "Matrilocality being a recurrent method of transcending the limited capacity of patrilineal village groups to form multi-village military alliances, it seems likely that societies on the threshold of statehood would frequently adopt matrilineal forms of social organisation" (92).

He cites Robert Briffault and several of the classical authors to suggest that many early European and Asian states had exhibited a matrilineal phase, a context in which marriage was matrilocal, women had relatively high status, and a cult of female ancestors was found. But this phase, as said, was short-lived, and few states, ancient or modern, have matrilineal kinship systems. As he puts it, "With the rise of the state, women again lost status . . . the old male supremacy complex reassert(ed) itself in full force" (1977, 93).

Although matrilineal kinship has virtually ceased to be a topic of interest among anthropologists (see Moore 1988, Ingold 1994), it has been of central concern to many Afro-centric scholars (Diop 1989) and eco-feminists, who have offered us lyrical accounts of a universal egalitarian "matriarchy" that existed prior to patriarchy and to the formation of the city-state, which is linked to the incursions of nomadic pastoralists from the Eurasian steppe. Given that matrilineal kinship is closely linked, as Harris suggests, with the rise of chieftains, I shall conclude this essay by critically discussing this literature.

Matriliny and Mother Goddess Religion

The notion that "matriarchy" was an original form of social organisation was a central doctrine of many early anthropologists. The writings of Jacob Bachofen (1967) on classical mythology and religion were particularly influential. Bachofen suggested that "all civilisation and culture are essentially grounded in the establishment and adornment of the hearth," and that "matriarchy" was an intermediate cultural stage in the development of human society, between hunter-gathering and the rise of the city-state. It was associated with the development of agriculture, mother-right (which did not necessarily imply the political domination of women), reciprocal rather than a Promethean attitude towards nature, and a religious system that emphasised human's dependence on the earth, expressed through chthonic deities. But although Bachofen suggested that at this stage of human evolution women were "the repository of all culture," he also emphasised that in all the classical civilisations—Egypt, Greece, Rome—an intrinsic relationship existed between "phallic" gods like Osiris—associated with water as a fecundating element—and female deities like Isis which were equated in myth with the earth, even though the latter were given prominence. Whenever we encounter matriarchy, Bachofen writes, we find it bound up with "chthonian religions," focused around female deities (88). He also makes the interesting observation that whereas the transience of material life goes hand-in-hand with matrilineal kinship, "father-right" is bound up with the immortality of a supramaterial life belonging to the "regions of light." With the development of patriarchy in the classical civilisations of Egypt and Greece, "the creative principle is dissociated from earthly matter," and comes to be associated with such deities as the Olympian gods (129). With the "triumph of paternity," humans are seen as breaking the "bonds of tellurism" (earthly life), and spiritual life rises over "corporeal existence." The "progress," as Bachofen views it, from matriarchy to patriarchy is thus seen by him as an important "turning point" in the history of gender relations (109).

The writings of Bachofen have had an enormous influence. Engels considered his discovery of matrilineal kinship—the original "mother-right gens"—as a crucial "stage" in human evolution; as on par with Darwin's theories in biology. In an often quoted phrase Engels suggested "the overthrow of mother right was the world historic defeat of the female sex" (1968, 488). Feminist anthropologists who have been influenced by Engels—such as Reed, Leacock, and Sacks—have thus strongly argued against the idea that the subordination of women is universal. They suggest that women have been significant producers in virtually all human societies, and that in many societies—particularly matrilineal societies—women have shared power and authority with men. Their activities were not necessarily devalued, and women often had a good deal of social autonomy, that is, they had decision-making power over their own lives and activities (Sacks 1979, 65–95; Leacock 1981, 134).

Anthropological and historical studies in recent decades have indicated the complexity and diversity of human cultures, and they have questioned whether "matriarchy" (however conceived) can be viewed simply as a "cultural stage" in the evolution of human societies. Yet in various ways Bachofen's bipolar conception of human history still has currency. For example, Bachofen has an unmistakable presence in the writings of the Senegalese scholar Cheikh Anta Diop (1989), though Diop gives Bachofen's thesis a strange twist in giving it a geographical and racialist interpretation. Thus matriarchy is seen as having flourished only in the "south" (Africa), and has as its correlate a settled agrarian way of life, a territorial state, gender equality, burial of the dead, and an ethic of social collectivism.

Patriarchy in Africa is linked to the intrusions of Islam. For all his scholarship, and his attempt to provide a more authentic anthropology,

Diop's work hardly captures the complexity of the history and culture of either Africa or Eurasia. But here I want to focus on the writings of some eco-feminists, especially those who espouse the "wisdom of goddess spirituality" (Spretnak, 1991). They too present an update and reinterpretation of Bachofen's simplistic bipolar conception of human history.

Whereas early classical scholars—such as Bachofen, Harrison, and Murray—saw chthonic deities as co-existing with male deities associated with the sun or sky—Ra, Apollo, Zeus, Amun—and implied that the latter deities came to have primacy only with the development of patriarchy and state structures, many eco-feminists now see the goddess as a "cosmic mother," a universal deity existent in all cultures prior to patriarchy. The male deities seem to be identified not with state structures—for mother goddess cults find their apotheosis in the theocratic states of Egypt and Crete—but with a later period of history with the emergence of imperial states and/or capitalism. Mother goddess cults are thus seen as a universal phenomenon, an expression of "ancient women's cultures" that once existed everywhere (Sjöö and Mor, 1987, 27).

While the proponents of the hunting hypothesis, like Ardrey (1976) suggest that all aspects of human life—language, intelligence, sociality, and culture—are derived from the "hunting way of life," with eco-feminists we now have the exact antithesis of this, and it is suggested that cultural life is essentially the creation of women. As Sjöö and Mor exclaim, "women created most of early human culture" (1987, 33). Unlike both Ardrey and these eco-feminists, perhaps early human communities were not obsessed with the gender division, and it is therefore probable that most basic life-tasks were shared, and thus human culture is the creation of both men and women.

Unlike Bachofen, who emphasised the "materiality" of matriarchy—based as it was on organic life—and thus associated "spirituality" with patriarchy, contemporary eco-feminists reverse this distinction and loudly proclaim the "spirituality" of matriarchy.

Aware, however, that there seems to be no historical evidence for matriarchy (as rule by women), feminist scholars have used terms like "communal matrifocal systems" or "matristic" to describe the more or less egalitarian communities that existed in the Palaeolithic (hunter-gathering) and Neolithic (agriculture) periods. Generally speaking, eco-feminists have tended to ignore anthropology, and have focused more on archaeology and classical studies, especially on mythology. They have, therefore, like Diop, presented us with a highly simplistic bipolar conception of human history. The latter is described in terms of an opposition between "ancient matriarchies" and a patriarchal system centred on men. We have the same kind of Gnostic dualism that Diop presented in his postulate of two "cradles" of humanity. Sjöö and Mor (1987) present this dualism with cogency, and it may be summarised as follows:

Ancient matriarchies	Modern patriarchy
religion based on deities associated with mother/earth	religion based on male deities
gender equality	gender hierarchy
partnership	domination
no sexual jealousy	sexual jealousy
harmony with nature	control over nature
matrifocal kin group	nuclear family
communalism	private property
holism	individualism
cyclic conception of time	linear conception of time
nurturing	greed and violence
chalice	blade

What is of interest, however, is that although Diop equated "matriarchy" with Black Africa, and even implied that the Dravidians of India were "Black Africans"—many classical scholars seem to follow their Victorian forebears in conflating race, culture, and language—contemporary ecofeminists see the historical dialectic between the "two social systems" as occurring within the European context itself. Sjöö and Mor's account of the "ancient religion" of the mother goddess largely focuses on Europe and on the cultures of classical antiquity—Egypt, Greece, Crete, and Sumeria. Riane Eisler's theory of cultural evolution, expressed in the readable *The Chalice and the Blade* (1987), focuses almost entirely on the European context and makes no mention of Africa at all.

Eisler's thesis is fairly straightforward and represents an elaboration and popularisation of ideas put forward long ago by Bachofen and Harrison. This suggests that the cultures of "old" or "ancient" Europe were based on settled agriculture, and were matrifocal, peaceful, ecocentric and focused on "mother goddess" cults which emphasised the life-generating and nurturing powers of the universe. Gender equality was the norm. It was symbolised by the chalice, the drinking cup. This "golden age" of female-oriented society that existed in "old Europe" (which Diop had argued was based on pastoralism and patriarchy) was either slowly transformed, or suddenly shattered—according to the archaeologist Marija Gimbutas (1974)—by marauding pastoralists migrating from the Asian steppes around 4000 BC, or patriarchy was facilitated by the rise of a military dictatorship, as in Babylon and Egypt (as Sjöö and Mor contend, 1987, 253). Whichever theory is embraced, there is the contention that European neolithic culture was radically transformed, from a

peaceful, sedentary, egalitarian, matrilineal society to one based on patriarchy. There was a "patriarchal shift" in old Europe, and the patriarchal
society which emerged was based on pastoralism, with its warrior ethic. Its
socio-cultural correlates were: the worship of male sky gods, the desacrali-
sation of the natural world, and an attitude of domination towards nature,
gender and social hierarchy, private property, and the state. In this process
the mother goddess cults were suppressed. This transition, according to
Eisler, represents a "cataclysmic turning point" in European history, and
the new patriarchal culture that emerged is symbolised by the blade—the
sword. A society based on partnership between men and women gave way
to one based on domination—including the domination of women by men.
Eisler presents this as a "new theory" of cultural evolution. But it is hardly
new: it is a Eurocentric re-statement of the theory of Bachofen and Engels.
Yet when we examine the ethnographic record concerning the religion of
hunter-gatherers, or even some small-scale horticultural societies, neither
matrilineal kinship nor mother goddess cults loom large. The religious
ideology of the Khoisan hunter-gatherers of Southern Africa and of the
Australian Aborigines hardly offers much support for the universality of
"mother goddess" forms of spirituality. Although there is a close identification with the natural world, particularly with animals, through totemic
spirits, or through spirits of the dead, there is little evidence among foragers of the deification of the earth itself as female, still less of the whole
universe. Equally, although there is a matrifocal emphasis among many
hunter-gatherers, as I discussed in my study of Hill Pandaram of South
India (1982), there is little emphasis on descent groups, and the key social
groups are the family and band. Kin groups may have salience for ritual
or marriage purposes, and may have totemic significance, but often, as
with the Australian Aboriginals, these are as likely to be patrilineal as
matrilineal. Among small-scale horticulturists in Melanesia and Amazonia,
as we noted above, patrilineal kinship has ideological stress, raiding and
homicide are endemic, and male initiation put a focal emphasis on the
training of young boys to be fierce warriors and to dominate women. Mary
Mellor (1992, 141–50) has drawn on this ethnographic material to question the assumption that clan-based or pre-state societies are necessarily
peaceful, or exhibit gender equality. Even matriliny, she remarked, was "no
guarantee against male violence" (47).

There is a unwarranted assumption among many feminist scholars
that matrilineal kinship, gender equality and mother goddess cults go
together, and necessarily entail each other. But the evidence suggests that
this is not so. For what is of interest is that cults focused on the "mother
goddess" and on the "earth-mother" find their richest elaboration not
among hunter-gatherers, nor among small-scale horticulturists, nor

indeed among societies that have a focal emphasis on matrilineal kinship—like the Iroquois and Bemba—but rather among theocratic states based on advanced agriculture, as Bachofen suggested. In an important survey of politics and gender among hunter-gatherers and small-scale horticulturists, Collier and Rosaldo (1981), much to their surprise, found little ritual celebration of women as nurturers or of women's unique capacity to give birth. Motherhood always formed a natural source of emotional satisfaction among women, and was culturally valued, but among such people fertility was not emphasised, and the deification of the "mother" as source of all life was generally absent. It is where you have complex states, where you have divine rulers like the Pharaoh and the Inca, who incarnate deities associated with the sun, that you find the earth deified, and motherhood ritually emphasised. For it was precisely among such theocratic societies based on intensive agriculture that there was a necessary emphasis on the land and on the reproduction of the labour force. Babylon and Egypt were not egalitarian "gardens of Eden" to the nomadic Hebrew pastoralists, but places where they were enslaved and subject to forced labour.

In an important sense, then, the deification of the earth as female and the emphasis on fertility—both of the land and of women—is a central tenet not of matrilineal societies like the Iroquois but of the patriarchal ideology of theocratic states. This ideology was clearly expressed in the writings of Francis Bacon, who identified women with nature, and advocated the knowledge and domination of both. Sherry Ortner, indeed, in a famous essay (1974), suggests an explanation for supposedly universal male dominance (patriarchy) by linking such dominance to an ideology that equates women with nature. For Ortner, then, "mother goddess" cults are a reflection of patriarchy, not of a matricentric culture. One feminist anthropologist has indeed argued that the "myth of matriarchy" is a fiction, and is used as a tool to keep woman "bound to her place" (Bamberger, 1974).

When we thus examine the early theocratic states of Crete and Egypt (for example), which are alleged to be matricentric paradises that exhibited gender equality and a peaceful social environment, what do we find? According to Janet Biehl (1991), what we find are highly developed bronze-age civilisations which, like theocratic states, were hierarchical, exploitative and oppressive. The theory of Gimbutas—that hierarchy emerged when a group of pure pastoralists arrived out of the Eurasian steppe and conquered pristine neolithic farmers—Biehl argues, is a naive simplification of European history, and scholars like Renfrew and Mallory would seem to agree (Biehl 1991, 43; Renfrew 1987, 95–97; Mallory 1989, 183–85).

Gender equality with regard to property, as in Egypt, may well have been restricted to the political elite; but in any case it co-existed, as Biehl points out, with an extremely hierarchical social structure focused around the pharaoh and a vast theocracy. The expansionist warfare, capital punishment and ritual sacrifices that were characteristic of most of these theocratic states—both in the Fertile Crescent and in the Americas—is generally overlooked or even dismissed by eco-feminist scholars. In the same way, Diop is an apologist for African state systems and the caste system as a form of social organisation.

Matriarchy has two distinct "foci" of meaning, which Bachofen tended to conflate. One is its connection with chthonic deities that associate the earth with motherhood; the other is with matrilineal kinship, which is a social group or category whose membership is determined by links through the female line. In social terms, the two meanings are not coterminous, for whereas mother goddess cults are associated with theocratic states and advanced agriculture, matrilineal kinship is associated with horticultural societies that lack both domestic animals and plough agriculture. Out of 564 societies recorded in the World Ethnographic Survey, David Aberle found only 84 (15 percent) where matriliny was the predominant form of kinship. He thus thought matriliny a "relatively rare phenomena" (1961, 663). Contrary to Diop's theory, matrilineal kinship is found throughout the world, but it is mainly found among horticultural societies that have developed chiefdoms. It is not found where there is intensive agriculture, or generally among pastoralists, or where state structures have developed—for patriarchy is intrinsically bound up with the state. Bachofen was of the belief that matriarchy was "fully consonant" with a situation where hunting, trade, and external raiding filled the life of men, keeping them for long periods away from women, who thus became primarily responsible for the household and for agriculture. Thus one may conclude that matriliny—but not mother goddess cults—seems to be particularly associated with horticultural societies that lack the plough, in which one finds developed political systems in the form of chiefdoms, and where there is what Poewe (1981) described as a complementary dualism between men and women. In these situations, subsistence agriculture is the domain of women, and men are actively engaged in hunting and trade that takes them for long periods away from the home base. Given their dominance in the subsistence sphere, women are not necessarily excluded from the public domain, and may be actively involved in public rituals and political decision-making. All the classical matrilineal societies that have been described by anthropologists essentially follow this pattern—the Bemba, Yao, and Luapula of Central Africa, the Trobriand Islanders, the Ashanti of Ghana, the Iroquois and Ojibwa of North America. All express a high degree of gender equality, sexuality is positively valued and there is an emphasis on sharing and reciprocity, but significantly there is little evidence of "mother goddess" cults. Such cults are bound up with the state and hierarchy, which is why they continued to flourish as an intrinsic part of Latin Christianity and Hinduism. There seems indeed to be a close correlation, as Harris suggests, between gender equality, matrilineal kinship, and the emergence of chiefdoms among horticultural societies.

References

Aberle, David. 1961. "Matrilineal Descent in Cross-Cultural Perspective." In *Matrilineal Kinship*, edited by David M. Schneider and Kathleen Gough, 655–80. Berkeley: University of California Press.

Ardrey, Robert. 1976. *The Hunting Hypothesis*. New York: Atheneum.

Bachofen, Johan. 1967. *Myth, Religion and Mother Right* (orig. 1861). Princeton, NJ: Princeton University Press.

Bailey, Frederick George. 1969. *Strategems and Spoils*. New York: Schocken Books.

Bamberger, Joan. 1974. "The Myth of Matriarchy." In *Women, Culture and Society*, edited by Michelle Zimbalist Rosaldo and Louise Lamphere, 263–80. Stanford: Stanford University Press.

Barclay, Harold. 1982. *People without Government*. London: Kahn and Averill.

Beidelman, Thomas O., ed. 1971. *The Translation of Culture*. London: Tavistock.

Biehl, Janet. 1991. *Rethinking Ecofeminist Politics*. Boston: South End Press.

Briggs, Jean L. 1970. *Never in Anger*. Cambridge, Massachusetts: Harvard University Press.

Brown, Judith K. 1975. "Iroquois Women." In *Toward an Anthropology of Women*, edited by Rayna R. Reiter, 235–51. New York: Monthly Review Press.

Carrithers, Michael. 1992. *Why Humans Have Cultures*. Oxford: Oxford University Press.

Chagnon, Napoleon. 1968. *Yanomamo: The Fierce People*. New York: Holt, Rinehart and Winston.

Clastres, Pierre. 1977. *Society against the State*. Oxford: Blackwell.

Cohen, Abner. 1974. *Two-Dimensional Man*. London: Routledge and Kegan Paul.

Collier, Jane F., and Michelle Z. Rosaldo. 1981. "Politics and Gender in Simple Societies." In *Sexual Meanings: The Cultural Construction of Gender and Sexuality*, edited by Sherry Ortner and Harriet Whitehead. Cambridge: Cambridge University Press.

Dahlberg, Frances, ed. 1981. *Woman the Gatherer*. New Haven: Yale University Press.

Dentan, Robert Knox. 1968. *The Semai: Non-Violent People of Malaya*. New York: Holt, Rinehart.

Diop, Cheikh Anta. 1989. *The Cultural Unity of Black Africa*. London: Karnak Press.

Ehrenreich, Barbara. 1997. *Blood Rites*. London: Virago Press.

Eisler, Riane. 1987. *The Chalice and the Blade*. London: Harper Collins.

Engels, Friedrich. 1968. "The Origins of Family, Private Property and the State." In Karl Marx and Friedrich Engels. *Selected Works*, 461–583. London: Lawrence and Wishart.

Fortes, Meyer, and E.E. Evans-Pritchard, eds. 1940. *African Political Systems*. Oxford: Oxford University Press.

Fried, Morton H. 1967. *The Evolution of Political Society*. New York: Random House.

Gibson, Thomas. 1989. "Symbolic Representations of Tranquility and Aggression Among the Buid." In *Societies at Peace: Anthropological Perspectives*, edited by Signe Howell and Roy G. Willis, 60–78. London: Routledge.

Gimbutas, Marija. 1974. *The Goddesses and Gods of Old Europe*. London: Thames and Hudson.

Gledhill, John. 1994. *Power and Its Disguises*. London: Pluto Press.

Harris, Marvin. 1977. *Cannibals and Kings*. London: Fontana.

———. 1993. "The Evolution of Human Gender Hierarchies." In *Sex and Gender Hierarchies*, edited by Barbara D. Miller, 57–59. Cambridge: Cambridge University Press.

Herdt, Gilbert H. 1987. *The Sambia*. New York: Holt, Rinehart.

Heywood, Andrew. 1994. *Political Ideas and Concepts*. London: Macmillan.

Hobbes, Thomas. 1651. *Leviathan* (1962 edition).

Howell, Signe. 1989. "To Be Angry Is Not to Be Human." In *Societies at Peace: Anthropological Perspectives*, edited by Signe Howell and Roy G. Willis, 45–59. London: Routledge.

Howell, Signe, and Roy G. Willis, eds. 1989. *Societies at Peace: Anthropological Perspectives*. London: Routledge.

Ingold, Tim. 1994. *Companion Encyclopedia of Anthropology*. London: Routledge.

Kelly, Raymond C. 1985. *The Nuer Conquest*. Ann Arbor: University of Michigan Press.

Kent, Susan. 1993. "Sharing in an Egalitarian Kalahari Community." *Man* 28, no. 3: 479–514

Knauft, Bruce M. 1987. "Reconsidering Violence in Simple Societies." *Current Anthropologies* 28: 457–500.

Kuper, Adam. 1988. *The Invention of Primitive Society*. London: Routledge.

Leacock, Eleanor B. 1981. *Myths of Male Dominance*. New York: Monthly Review Press.

———. 1983. "Ideologies of Male Dominance." In *Woman's Nature: Rationalizations of Inequality*, edited by Marian Lowe and Ruth Hubbard, 111–21. New York: Pergamon Press.

Lee, Richard B. 1979. *The !Kung San*. New York: Cambridge University Press.

Levy, Robert I. 1973. *Tahitians*. Chicago: University of Chicago Press.

Lewellen, Ted C. 1992. *Political Anthropology*. Westport: Bergin and Garvey.

Lizot, Jacques. 1985. *Tales of the Yanomami*. Cambridge University Press.

MacPherson, C.B. 1962. *The Political Theory of Possessive Individualism*. Oxford: Clarendon Press.

Mallory, J.P. 1989. *In Search of Indo-Europeans*. London: Thames and Hudson.

Maximoff, G.P., ed, 1953. *The Political Philosophy of Bakunin*. Glencoe, IL: Free Press.

Maybury-Lewis, David. 1971. *Akwe-Shavante Society*. Oxford: Clarendon Press.

Mellor, Mary. 1992. *Breaking the Boundaries*. London: Virago.

Middleton, John, and David Tait, eds. 1958. *Tribes without Rulers*. London: Routledge and Kegan Paul.

Mill, John Stuart. 1972. *Utilitarianism, On Liberty*. London: Dent.

Moore, Henrietta L. 1988. *Feminism and Anthropology*. Oxford: Polity Press.

Morris, Brian. 1982. *Forest Traders*. London: Athlone Press.

Nance, John. 1975. *The Gentle Tasaday*. New York: Harcourt Brace.

Ortner, Sherry B. 1974. "Is Female to Male as Nature Is to Culture?" In *Women, Culture and Society*, edited by Michelle Zimbalist Rosaldo and Louise Lamphere, 67–88. Stanford: Stanford University Press.

Overing, Joanna. 1989. "Styles of Manhood: An Amazonian Contrast in Tranquillity and Violence." In *Societies at Peace: Anthropological Perspectives*, edited by Signe Howell and Roy G. Willis, 79–99. London: Routledge.

Poewe, Karla. 1981. *Matrilineal Ideology*. London: Academic Press.

Radcliffe-Brown, A.R. 1940. Preface. In *African Political Systems*, edited by Meyer Fortes and E.E Evans-Pritchard. London: Oxford University Press.

Renfrew, Colin. 1987. *Archaeology and Language*. Harmondsworth: Penguin Books.

Robarchek, Clayton A. 1989. "Hobbesian and Rousseauan Images of Man." In *Societies at Peace: Anthropological Perspectives*, edited by Signe Howell and Roy G. Willis, 31–44. London: Routledge.

Sacks, Karen. 1979. *Sisters and Wives*. Urbana: University of Illinois Press.

Service, Elman R. 1962. *Primitive Social Organization*. New York: Random House.

Shostak, Marjorie. 1981. *Nisa: Life and Words of a !Kung Woman*. Cambridge: Harvard University Press.

Silberbauer, George. 1982. "Political Process in G/wi Bands." In *Politics and History in Band Societies*, edited by Eleanor Leacock and Richard B. Lee, 23–36. Cambridge: Cambridge University Press.

Sjöö, Monica, and Barbara Mor. 1987. *The Great Cosmic Mother*. New York: Harper and Row.

Spiro, Melford E. 1952. "Ghosts, Ifaluk, and Teleological Functionalism." *American Anthropology* 54: 497–503.

Spretnak, Charlene. 1991. *States of Grace*. New York: Harper Collins.

Thomas, Elizabeth Marshall. 1958. *The Harmless People*. New York: Random House.

Turnbull, Colin M. 1961. *The Forest People*. New York: Doubleday.

_____. 1982. "The Ritualisation of Potential Conflict Between the Sexes Among the Mbuti." In *Politics and History in Band Societies*, edited by Eleanor Leacock and Richard B. Lee, 133–55. Cambridge University Press.

Weber, Max. 1947. *The Theory of Social and Economic Organization*. New York: Free Press.

Woodburn. James. 1982. "Egalitarian Societies." *Man* 17: 431–51.

9

Reflections on the "New Anarchism" (2008)

Prologue

No doubt you have heard about the coming "new age" and all about "New Labour." No doubt you have also read lots about postmodernism, post-structuralism, postfeminism, post-Marxism, and posthumanism. There are even postanimals around, but they are not to be confused with the real badgers and dormice that inhabit the woods and fields. So you will not be surprised to learn that academic scholars have now discovered what is described as the "new anarchism" along with one of its variants, "post-anarchism" (Day 2005, Kinna 2005, Curran 2006).

The suggestion is that anarchism as normally understood has become an "historical baggage" that needs to be rejected, or at least, given a "major overhaul" (Purkis and Bowen 1997, 3).

The anarchism of an earlier generation of anarchists is thus declared to be "old anarchism" and is perceived to be old-fashioned and outdated; or as John Moore put it in the pages of the *Green Anarchist,* just plain "obsolete," a historical relic of no relevance at all to contemporary radical activists (Holloway 2002, 21).

Embracing a crude linear, bipolar mode of understanding the history of anarchism—an approach that is facile, undialectical, and lacking any sense of history—we are told by the academics that "old anarchism" is now antiquated and "outmoded" (Kinna 2005, 21). By "old" anarchism they essentially mean social anarchism or class struggle anarchism (mutu-alism, libertarian socialism, anarcho-syndicalism, anarchist commu-nism)—which are, unbeknown to these academics, still the most vibrant strands of anarchism around, judging by recent texts (Sheehan 2003, Franks 2006).

Nevertheless, we are informed that "old" anarchism has been replaced by a new variety of anarchism—the "new anarchism." In fact, there has been a "paradigm shift" (no less!) within anarchism itself (Purkis and Bowen 2004, 5).

This "new anarchism" as the "new paradigm" appears to be an esoteric pastiche of poetic terrorism (otherwise known as Nietzschean aesthetic nihilism), anarcho-primitivism, the radical individualism (egoism) of Max Stirner, and an appeal to the oracular musings of post-structural philosophers such as Jean-François Lyotard, Michel Foucault, Jacques Derrida, Gilles Deleuze, and Jean Baudrillard. None of whom, it is worth noting, were anarchists.

We are also joyfully informed that no contemporary radical activist has ever read the works of Bakunin, Kropotkin, or Maletesta, as they are deemed to be as old-fashioned as the novels of Charles Dickens (Purkis and Bowen 1997, 3). This is probably true. For the obvious reason that few people in the new social movements or in the recent anti-globalisation protests are in fact anarchists.

1. New Social Movements

Take the new social movements. Anarcho-feminists were in a distinct minority in the second-wave feminist movement. Most feminists were liberals, Marxists or republicans, embracing an identity politics that appealed to state power to enact reforms. There were also few anarchists in the American civil rights movement, though this movement was well represented by black nationalists and radical liberal pacifists.

And the ecology movement as I long ago indicated (1996, 131) embraces people right across the political spectrum. It includes liberals like Jonathon Porritt (now a staunch advocate of capitalism), members of the Green Party and other worthy liberals, outright authoritarian conservatives such as William Ophuls, Garrett Hardin, and Rudolf Bahro, as well as German neo-fascists and the followers of the Nazi sympathiser Martin Heidegger, the darling of some deep ecologists and postmodernists. Within the ecology movement anarchists are therefore in a distinct minority, even though anarchism is the only political tradition that is fully consonant with an ecological sensibility—as Bookchin (1982) long ago argued (Morris 1996; Hay 2002, 280–97).

Anti-Globalisation Protests

Equally, anarchists form a minority in the anti-globalisation protests, although their presence is invariably highlighted by the media, especially when they destroy property. Most of the radical activists on the anti-globalisation protests are reformist liberals who simply seek to humanise capitalism and make it more benign. Some like the late Pierre Bourdieu (1998) and the Brazilian Workers' Party (who hosted the first World Social Forum in Porto Alegre) merely want to bolster the economic power of the nation-state, and thus curb the worst excesses of global capitalism. Some like Susan George and the French organisation Association for the Taxation of Financial Transactions to Aid Citizens (ATTAC) simply advocate putting a tax on the movement of capital. Others still, like David Held, Arne Naess, and George Monbiot—who has taken over Jonathon Porritt's mantle as the media radical on environmental issues—envisage some kind of "global democratic state"—heaven forbid!

It is of interest that George Monbiot is always complaining about the "antics" of anarchists on the anti-globalisation protests—for the anarchists lack discipline, destroy property, and upset the police. Like a worthy liberal Monbiot seeks to uphold the sanctity of private property, and views the police as the benign custodians of "law and order" (see Monbiot 2000).

To equate the anti-globalisation protests with anarchism is therefore quite misleading. Even so, anarchists have made their presence felt—through the Black Bloc and with the politics of détournement—involving guerilla advertising and an emphasis on the aesthetic dimensions of protest. Such forms of protest are hardly novel. At the end of the eighteenth century Thomas Spence called his weekly periodical *Pig's Meat, or Lessons for a Swinish Multitude*, precisely to parody Edmund Burke's derogatory opinion of working people. Détournement and symbolic protest did not begin with the Situationists; and even the concept of "multitude" is hardly a new idea (on Spence's agrarian socialism see Morris 1996, 112–22).

This is not to deny, of course, that there have been important anarchist tendencies in both the new social movements and the anti-globalisation protests. These have been of particular interest to academic scholars (Day 2005, Curran 2006). One writer indeed suggests that the "soul" of the anti-capitalist movement is anarchist, in its disavowal of political parties, and its commitment to direct action. The movement is thus, he writes, firmly in the spirit of libertarian socialism—that is, the spirit of the "old anarchism" (Sheehan 2003, 12).

It is also of interest that in the wake of the anti-globalisation protests, Purkis and Bowen seem to have revised their opinion about the "old anarchists." For they write that contemporary radical activists are seeking out the writings and quoting from the works of Kropotkin, Proudhon, Bakunin, Goldman, and Malatesta—insisting that there are still issues and principles that are worthy of debate, despite a very different context (2004, 2). True.

2. The New Anarchists

The ideas and practices of contemporary mutualists and class struggle anarchists are, of course, just as "new" as those of primitivists, Stirnerite individualists and the self-proclaimed poetic terrorists. And certainly class struggle anarchists—those who marshal under such banners as Class War, the Solidarity Federation, the Anarchist-Syndicalist Network, and the Anarchist (Communist) Federation—have been just as much involved in radical anti-capitalist protest as have the likes of Hakim Bey, the Autonomist Marxists and the poststructuralists. Protest and radical activism have always been an essential part of class struggle anarchism (see Sheehan 2003, Franks 2006).

But who are these "new anarchists"? Well, according to Ruth Kinna, they essentially muster under six ideological categories, namely: Murray Bookchin's eco-socialism; the anarcho-primitivism of John Zerzan; the acolytes of the radical individualism of Max Stirner; the poetic terrorism (so-called) of Hakim Bey and John Moore, who follow the rantings of the reactionary philosopher Friedrich Nietzsche; the postmodern anarchism derived from Deleuze, Foucault, and Lyotard; and finally (believe it or not), the anarcho-capitalism of Murray Rothbard and Ayn Rand.

Although admitting that there has been some antipathy between these various strands of the "new anarchism," what they have in common, Kinna tells us, is that they have all repudiated the "struggle by workers for economic emancipation" (2005, 21). That is, they have abandoned libertarian socialism, or "leftism" (socialism), to use the current derogatory label (Black 1997). This is, of course, analogous to the politics of New Labour, as both Nicolas Walter and Graham Purchase suggested in their review of *Twenty-First Century Anarchism* (Purkis and Bowen 1997).

Two points need to be made initially regarding Kinna's depiction of this "new anarchism."

Bookchin's Eco-Socialism

The first point is that Murray Bookchin would undoubtedly refute being described as an advocate of the "new anarchism," or what he himself called "lifestyle" anarchism. Although Bookchin, like Bakunin and Kropotkin before him, certainly did not envisage the "industrial proletariat" as the sole revolutionary agent, he never repudiated class struggle anarchism. In his last years he may indeed have jettisoned the label "anarchist," as in the United States, anarchism had become virtually synonymous with anarcho-primitivism and aesthetic nihilism. And his strident advocacy of municipal socialism also found little favour among anarcho-communists and other class struggle anarchists. Nevertheless, Bookchin always acknowledged the need for serious class analysis, and affirmed the crucial importance of working class struggles in achieving any form of social revolution. What he attempted to do with his concept of hierarchy was to broaden existing conceptions of social oppression. Bookchin, therefore, always remained a revolutionary libertarian socialist, and had little but disdain for poetic terrorism, primitivism, technophobia, mysticism, and anti-rationalism of the "new" or lifestyle anarchists. He was equally critical of the relativism, incoherence, and nihilism of postmodern philosophers, who, he felt, tended to denigrate reason, the objectivity of truth, and the reality of history (Bookchin 1995, 1999).

What is of interest is that Bookchin regarded Kropotkin as perhaps the most far-seeing of all the theorists he encountered in the libertarian tradition. Bookchin also emphasised—unlike the "new anarchists"—the need to respect an earlier generation of anarchists (Bakunin, Kropotkin, Reclus, Malatesta) not only for what they achieved in their own lifetime, but also for what they have to offer contemporary radical activists. But he also stressed the need to develop, build on, and go beyond their ideas, rather than arrogantly dismissing them as "obsolete" (Moore) or "irrelevant" to contemporary struggles (Holloway; Bookchin 1993, 55–57).

The powers and intrusions of the modern state have undoubtedly increased over the past century, while capitalism has so expanded that it has turned virtually everything into a commodity; nevertheless the practices and theoretical perspectives of the early class struggle anarchists like Bakunin and Kropotkin still have a contemporary relevance. (On the continuing relevance of Bakunin, Kropotkin, and Elisée Reclus see, for example, Morris 1993, 2004, McLaughlin 2002, Leier 2006, Clark and Martin 2013.)

Anarcho-Capitalism

A second point is that the anarcho-capitalists are not by any stretch of the imagination anarchists—as this term is normally understood. Take Ayn Rand. Her political philosophy Ruth Kinna turns into another "ism," Aynarchism—adding yet another "ism" to the thirty or so that adorn her introductory chapter "What Is Anarchism" (enough to bamboozle any "beginner" to the subject!) (2005, 25). But Rand, an early devotee of Nietzsche, explicitly repudiated anarchism, advocated a minimal but highly repressive state whose sole function was to support capitalist exploitation, and was a fervent promoter of free-market capitalism. She was in fact the intellectual guru of Margaret Thatcher. An egoist in ethical theory, anti-feminist and anti-ecology, Ayn Rand saw city skyscrapers as a positive symbol of American capitalism, and of the human conquest of nature. Her vision is thus the exact antithesis of John Zerzan's anarcho-primitivism. How on earth scholars like Kinna (2005, 35) and Simon Tormey (2004, 119) can consider Ayn Rand as "anarchist" is beyond my comprehension. She was essentially an elitist, liberal republican (on Rand's political philosophy see Morris 1996, 183–92).

3. What's New in the New Anarchism?
What is of interest and significant is that very little of this "new anarchism" is in fact new or original. Let us discuss each strand in turn.

Anarcho-Capitalism
"Aynarchism," for a start, is just a re-affirmation of nineteenth-century laissez-faire capitalism. The "egoism" embraced by Ayn Rand is, in fact, the very apotheosis of bourgeois thoughts on the individual, which go back to the classical liberalism of Hobbes and Locke.

Stirner's Individualism
The radical individualism advocated by the acolytes of Max Stirner is also somewhat antiquated. With Stirner, whose egoism is uncritically embraced by both Hakim Bey (1991) and John Moore (2004), the self becomes its own "property" and taking on the state form feels free to exploit and dominate others. "L'état, c'est moi," the state, it is me, Hakim Bey proclaims (1991, 67). This, surely, is not an anarchist sentiment, and is hardly conducive to mutual aid and voluntary cooperation.

Stirner was, of course, a left-Hegelian (not a poet!) and was critiqued by Kropotkin at the end of the nineteenth century (Baldwin 1927, 161–72).

Although acknowledging Stirner's historical importance, Kropotkin considered his amoral egoism limited and stultifying, in that Stirner repudiated neither property nor the state in his sanctification of the unique "ego." As Stirner puts it, "I do not want the liberty of men, nor their equality; I want only *my* power over them. I want to make them my property, *i.e.*, *material for my enjoyment*" (1973, 318). Humans are thus not to be respected as persons, but seen only as an "object" for the ego's enjoyment (311).

Interestingly, the anarchist critique of Stirner's egoism is completely ignored by his recent admiring devotees—Hakim Bey, in fact, tries to convince us that Stirner was not an "individualist" but embraced a joyous "conviviality" (1991, 67–71). Stirner certainly acknowledged an "association" of egoists, but as John P. Clark argued long ago, Stirner has little understanding of such values as community, solidarity, cooperation, and mutual aid because he has such an abstract conception of the human individual (1976, 97).

But, both Stirner's and Ayn Rand's egoism is merely an expression of bourgeois possessive individualism. To use a common expression, it is all "old hat." Although of course the "new" Stirnerite individualists like to appropriate the term "anarchy" for themselves, contrasting it with "old-fashioned" class struggle anarchism or "leftism" (Black 1997, Moore 2004; for a useful, succinct critique of Stirner's individualism see Bookchin 1999, 125–26).

It is then quite ironic, if not perverse, to see academics like Saul Newman (2002) interpreting Stirner as an anti-essentialist thinker, when in fact Stirner was an essentialist *par excellence*, and was lampooned and critiqued as such by Marx and Engels long ago (1846/1965). Stirner's concept of the human individual as possessive, power-seeking, exploitative, amoral, competitive, and atomistic is thoroughly abstract and Hobbesian (bourgeois), as Kropotkin indicated in critiquing Stirner's essentialism. Kropotkin, of course, like every other social anarchist, recognised that all humans have unique personalities; but articulating a social, non-essentialist conception of the human subject (unlike Stirner) Kropotkin stressed that the freedom, integrity, and self-development of the individual could only be achieved in a free society—what Kropotkin termed free communism.

Poetic Terrorism

The kind of radical "aestheticism" and "cultural elitism" that stems from Friedrich Nietzsche is also hardly "new." For it was fashionable among the avant-garde at the end of the nineteenth century. Now resurrected and labelled "poetic terrorism," "attentat art" (art as protest), "ontological anarchy" or "radical aristocratism" (take your pick!) it is again well represented by the "lifestyle" anarchism of Hakim Bey (1991) and John Moore (2004c). Indeed, Bey's writings represent an incoherent esoteric pastiche of the following: anarcho-primitivism (while completely oblivious to the mutual aid and communist ethos of tribal hunter-gatherers); the asocial egoism of Max Stirner; the "psychic nomadism" of the poststructuralists Deleuze and Guattari (while completely ignoring their radical materialism); the aesthetic elitism and "poetic terrorism" of Nietzsche; new-age spiritualism, including the joyful embrace of Islamic mysticism; chaos theory (misunderstood!); along with Proudhonian federalism, cyberspace, the use of Black Magic as a revolutionary tactic (while denying the possibility of social revolution!); and—watch out!—"a *psychic paleolithism* based on High-Tech" (1991, 43).

In contrast to philosophers like Hegel and Whitehead, Hakim Bey makes a fetish of incoherence (as does Moore) and expresses his thoughts in the most pretentious, scholastic gobbledygook, designed by both to impress and intimidate the ordinary reader (Moore tends to follow in his wake). No wonder Murray Bookchin and John Zerzan both dismiss Hakim Bey's mystical blather as elitist, petit bourgeois, and basically reformist.

Hakim Bey's concept of the "temporary autonomous zone" is likewise, hardly original, though it has provoked a bucket load of scholarly debate.

Over the past century anarchists, as well as ordinary people, have been involved in autonomous activities: some fleeting, some enduring. They have thus created trade unions, affinity groups, communes, cooperatives, voluntary associations, and anarchist organisations, all of which have been independent of both the state and the capitalist economy. Some are workplace organisations; some community based. They have thus long ago been engaged in the creation of autonomous "zones" of social activity. As well, of course, being involved in temporary events—sabotage, strikes, protests, and demonstrations. In fact, for generations of class struggle anarchists "spontaneity" has always implied not just transitory events but rather the creation of non-hierarchical forms of organisation that are truly organic, self-created, and voluntary—as Bookchin expressed it long ago in his pamphlet: "spontaneity and organization" (1972/1980, 251–74). Thus there is precious little that is new or original in Hakim Bey's conception of a "temporary autonomous zone." It is what Colin Ward (1973)—also long ago—called "anarchy in action."

What *is*, however, new about Bey's concept is that it is purely fleeting and ephemeral, and is combined with the notion of a "nomadic individual," the archetypal lone-ranger, who leaves to other mortals the care and upbringing of children, and the production of food and the other necessities of human life, all of which require some form of organisation. Ignoring forms of social oppression and widespread inequalities, indeed abandoning any form of resistance to social oppression, Bey's liberal politics confronts neither capitalism nor state power, but happily and joyfully co-exists with them. All the while, indulging in disinterested aesthetic contemplation, blissful and ludic, along with occult practices and Islamic mysticism. All of which are thoroughly reactionary, and can hardly be described as either anarchism or anarchy. In fact, Bey's eclectic postmodern pastiche sits rather well with New Age Spiritualism and consumer capitalism. The choice that Bey (1994) offers us, that between immediatism and capitalism is, of course, thoroughly facile. (For useful critiques of Bey's liberal politics see Bookchin 1995b, 20–26; Zerzan 2002, 144–46; Franks 2006, 266–67.)

Contrary to what Giorel Curran (2006) suggests, the concept of TAZ, or "temporary autonomous zone," has very little connection with the "carnivals of rebellion" or the "festivals of resistance" that have been part and parcel of such movements as Reclaim the Streets, the anti-roads movement and the anti-globalisation protests. Indeed street carnivals and festivals have always been an integral part of class struggle, anarchism, and revolutionary socialist movements ever since the French Revolution.

Recent attempts by academic scholars to interpret Nietzsche as an "anarchist" (no less!) (Moore 2004b; Sheehan 2003, 72) seem to me quite fallacious. Although there is undoubtedly a libertarian aspect to Nietzsche's philosophy—for example, his solitary form of individualism with its aesthetic appeal to self-making; the radical critique implied in his "revaluation of all values"; his strident attack on the state in *Thus Spake Zarathustra*; and his impassioned celebration of the life-instincts and personal freedom and power. This is why Nietzsche had such an appeal to Emma Goldman and Guy Aldred.

But all this is more than offset by Nietzsche's thoroughly reactionary mindset. This is illustrated by his elitist politics, his celebration of authority and tradition, his misogyny, his admiration of the Indian caste-system and dictators like Julius Caesar and Napoleon, and his complete lack of any progressive vision apart from the notion of an isolated, asocial nomad, the "overman" who will be the legislator of humankind. Reciprocity, mutual aid, and equal rights for all were "poisonous" doctrines according to Nietzsche, for what he valued was a "good and healthy aristocracy" (his words). Nietzsche was, indeed, in many respects, as Richard Wolin argues (2004, 52–63), a proto-fascist (see my critical review, Morris 2007).

It is worth noting that although Kropotkin admired Nietzsche's poetic writings, and thought he was unequalled in his critique of Christianity, and was "great" as a theorist of "revolt," he nevertheless felt that Nietzsche always remained a "slave to bourgeois prejudices." For Nietzsche had little understanding of either socialism or anarchism, and his philosophy was not so much a repudiation or "refusal" of "modernity" as its apotheosis (Kropotkin 1970, 305).

Primitivism

Turning now to anarcho-primitivism, this too is hardly "new." Primitivism is, of course, as old as the hills, and goes back to the beginnings of agrarian civilisation. It is particularly associated with the Enlightenment philosopher Jean-Jacques Rousseau, and the concept of the "noble savage." John Zerzan (1988, 1994) and other anarcho-primitivists simply embrace this ancient idea, weld it to contemporary anarchism, and declare, in utopian fashion, that the "future" is primitive. This entails, one supposes, some kind of return to a tribal hunter-gathering existence?

John Zerzan presents an apocalyptic, even perhaps a gnostic vision. Our hunter-gatherer past is thus described by Zerzan as an idyllic era of virtue and authentic living. The last eight thousand years or so of human history after the rise of agriculture (the fall) is seen as a period of tyranny and hierarchical control, a mechanised routine devoid of any spontaneity, and involving the anaesthetisation of the senses. All those products of the human creative imagination—farming, art, philosophy, technology, science, writing, urban living, symbolic culture—are viewed negatively and denigrated by Zerzan in a monolithic fashion.

The future, we are told, is "primitive." How this is to be achieved in a world that presently sustains around six billion people (for evidence suggests that the hunter-gatherer lifestyle is only able to support one or two people per square mile) or whether the "future primitive" actually entails a return to hunter-gatherer subsistence Zerzan does not tell us. Whether such images of green primitivism are symptomatic of the estrangement of affluent urban dwellers and intellectuals from the natural (and human) world, as both Roy Ellen (1986) and Murray Bookchin (1995, 120–46) suggest, I will leave others to judge.

But what is important about Zerzan's work is the affirmation that hunter-gatherers were, in many respects, "stone age anarchists." As Zerzan puts it, life in "primitive" society was largely one of leisure, gender equality, intimacy with the natural world, and sensual wisdom. All this, of course, was emphasised by Peter Kropotkin long ago. Kropotkin stressed the close intimacy that existed between humans and animals in tribal society, that tribal people had an encyclopaedic knowledge of the natural world, and put a marked emphasis on sharing, generosity, and mutual aid, which co-existed with an equal emphasis on individual autonomy and independence. But, unlike Zerzan, Kropotkin was not blind to the limitations of tribal life, the oppressive nature of certain traditions and the hierarchical aspects of tribal life (Kropotkin 1902, 74–101; Morris 2004, 173–90).

When John Moore embraced primitivism as a source of inspiration for contemporary anarchism he was not, then, suggesting anything new or original. Kropotkin had suggested this a century earlier. Indeed, having undertaken ethnographic studies among hunter-gatherers living in the Ghat Mountains of South India, more than two decades before John Moore penned his "primitivist primer" and having experienced the reality of hunter-gatherer social existence, I personally never contemplated, any more than did Kropotkin, becoming a permanent forager. Thus we need to draw inspiration and lessons from the cultural past of tribal peoples, as Kropotkin suggested, without romanticising them, or trying to emulate the social life of the hunter-gatherers (Morris 1982, 1986, 2005, 4–6; for critiques of Zerzan's primitivism see Bookchin 1995, 39–49; Shephard 2003; Albert 2006, 178–84).

It is of interest that Giorel Curran (2006, 42) outlines some of the criticism of Zerzan's eco-primitivism—as being misanthropic, as fanciful and hopelessly romantic, and as dismissive of the human potential for creativity and innovation, without ever indicating the source of such criticisms or whether in fact she agrees with them.

Postmodern Anarchism

Let me finally turn to so-called postmodern anarchism, otherwise known as poststructuralist anarchism, or simply post-anarchism.

In the last decades of the twentieth century the academy was suddenly besieged by the rhetoric of the so-called "postmodernists," who, following in the footsteps of Nietzsche, Heidegger, and Wittgenstein (all political reactionaries), have had a baneful influence on the social sciences. Now, somewhat belatedly, it seems that anarchists have also become infatuated with postmodernism—at a time when most social scientists are leaping off this particular bandwagon, and writing books like *After Postmodernism* (Lopez and Potter 2001).

Postmodernism is a diffuse, rather inchoate cultural movement or ideology that is difficult to define as it includes scholars with radically different approaches to social life. But as an intellectual ethos associated with such scholars as Baudrillard, Lyotard, Derrida, Foucault, Rorty, Laclau, and Butler, it has been characterised by the following tenets, all presented in the most oracular fashion.

Firstly, as (supposedly!) we have no knowledge of the world except through "descriptions" (to use Rorty's term) the "real" is conceived as an "effect" of discourses. Anti-realism is thus in vogue, and we are informed that there is no "objective reality." As the anthropologist Mary Douglas put it, "All reality is social reality" (1975, 5). Or as Derrida put it "there is nothing outside the text" (1976, 158). He was later to plead that he had been misunderstood (so meaning is not, as he always claimed, indeterminate!), and that he did not doubt the reality of the material world, but only wished to advance a "textualist" approach (Rötzer 1995, 45).

Postmodernism therefore tends to propound a neo-idealist metaphysic, and thus to repudiate *realism*—the notion that the natural world has an objective reality and existence independent of human consciousness.

Secondly, as our perceptions and experiences of the world are always to some degree socially mediated—acknowledged by generations of social scientists ever since Marx's reflections on Feuerbach's materialism—postmodernists take this insight to extreme and come to espouse an epistemological (and moral) relativism. Truth is either repudiated entirely (Tyler 1986), or seen simply as an "effect" of local cultural discourses (Rorty 1979, Flax 1990), or something that will be "disclosed" by elite scholars (Heidegger 1994). Truth as correspondence is thus repudiated, and as knowledge is always historically and socially situated, there can be, it is argued, no universal truths or values. Truth, we are told, is always subjective, indeterminate, relative, and contingent. Both objective knowledge and empirical science are thus repudiated, along with universal values.

Thirdly, critiquing the transcendental ego of Cartesian rationalism (and phenomenology) together with the abstract individual of bourgeois ideology—which had, of course, been critiqued, indeed lampooned, by Marx, Bakunin, and Kropotkin in the nineteenth century!—postmodernists

again go to extreme and announce the "dissolution" of human subjectivity, the self being declared a "fiction." Human agency is thus repudiated,
or the subject is viewed simply as an "effect" of "ideology" or "power" or
"discourses."

Fourthly, there is a rejection of "metanarratives" (Lyotard 1984)—
Marxism, Buddhism, human rights discourses, evolutionary biology, pale-
ontology, universal history—and a strident celebration expressed of the
"postmodern condition." The so-called postmodern condition—with its
alienation, cultural pastiche, fragmentation, decentred subjectivity, nihilism—describes, however, not so much a new paradigm or epoch, but rather
the cultural effects of global capitalism.

Finally, there has been a growing tendency among postmodern academics—following Heidegger—to express themselves in the most obscure
and impenetrable jargon, under the misguided impression that obscurity
connotes profundity, and that a neo-Baroque prose style is the hallmark of
radical politics. It isn't! (Morris 1997, 2006, 8–10; Hay 2002, 321–22).

All this has led the acolytes of postmodernism to proclaim, with some
stridency, the "dissolution" or "erasure" or the "end" of such concepts as
truth, reason, history, class, nature, the self, science, and philosophy—along
with the Enlightenment "project" itself. Yet in their rejection of history and
class, in reducing social reality to discourses, in their epistemic and moral
relativism, in their dissolution of human subjectivity, and in their seeming
obsession with the media, high-tech cyberspace, and consumer capitalism,
many have remarked that there seems to be an "unholy alliance" between
postmodernism and the capitalist triumphalism of the neoliberals (Wood
and Foster 1997).

Postmodernism has, of course, been subjected to a barrage of criticism from numerous scholars, which makes one wonder why it is still
held in such esteem by some academic anarchists. (For critiques, see, for
example, Gellner 1992, Bunge 1996, Callinicos 1997, Detmer 2003, and from
an anarchist perspective Bookchin 1995, 172–204; Zerzan 1994, 101–34.)

Poststructuralism is sometimes described as the philosophy of postmodernism. Yet what is of interest is that the radical scholars that have
most appeal to the post-anarchists—Foucault, Derrida, Deleuze, Guattari—
all expressed an opposition to postmodernism, or at least distanced themselves from it. Yet it also has to be recognised that few of the political ideas
expressed by the poststructuralist philosophers are in fact new or original.
For what they have done is simply to appropriate many of the basic ideas
and principles of (social) anarchism, and wrapped them up in the most
scholastic jargon—with little or no acknowledgement. These ideas, please
note, include the following: opposition to the state and all forms of power
and oppression; recognition that the power of the modern state intrudes

into all aspects of social life; a fervent anti-capitalism; a rejection of the vanguard party, representation and the notion that the transformation to socialism can be achieved through state power; a rejection of an "essentialist" (i.e., Cartesian/abstract) conception of human nature; and finally, the importance of creating alternative social forms of organisation, non-hierarchic and independent of both the state and capitalism.

Kropotkin (and other social and class struggle anarchists over the past century) had, of course, expressed these ideas long ago—more concretely, and much more lucidly. Unlike most poststructuralists, from Bourdieu to Baudrillard and Derrida, Kropotkin (and Elisée Reclus) also expressed an ecological sensibility.

Here is a typical extract from Deleuze, discussing French capitalism: "Against this global policy of power, we initiate localised counter-responses, skirmishes, active, and occasionally preventative defences. We have no need to totalise that which is invariably totalised on the side of power; if we were to move in this direction, it would mean restoring the representative forms of centralism and a hierarchical structure" (Foucault 1977, 212). Deleuze seems singularly unaware that this strategy had been advocated by Kropotkin and the anarchist tradition for more than a hundred years.

All this is well illustrated in Dave Morland's (2004) suggestions about the poststructuralist anarchism supposedly being expressed in recent new social movements. He writes that such anarchism repudiates representational politics, advocates "autonomous capacity building" (what Kropotkin and early anarchists described simply as direct action!), is suspicious of vanguardism and revolutionary elites, and shuns the quest for political power. This is not some "new mode" of anarchism, and there is nothing "poststructuralist" about it: it simply reflects what social anarchists have been advocating and practising for the last hundred years. Social or class struggle anarchists have in fact always been a constituent part of *all* protest movements since the Second World War, whether against fascism, nuclear power, the Vietnam War, the poll tax or, more recently, global capitalism. Class struggle anarchists have always been around, and were in evidence long before the anti-globalisation protests, when they received especial media attention as a "travelling circus" (Goaman 2004).

Likewise Saul Newman's definition of "postanarchism" simply regurgitates what anarchists like Kropotkin long ago emphasised: the repudiation of relations of domination; an ethic of mutual aid; the intrinsic relationship between equality and liberty; a commitment to respect "difference" (diversity) and individual autonomy within a collectivity; and an emphasis on community which is not equated with a "herd" mentality. But at least Newman, unlike his mentors, acknowledges his sources, namely the writings of Bakunin and Kropotkin. (2004, 123; see Morris

2004). Postanarchism is simply an exercise of putting old wine into new wine bottles! Richard Day (2005, 123) even describes Kropotkin as the "first postanarchist." Kropotkin wasn't "post" anything; he was an anarchist. He was part of an ongoing revolutionary movement and political tradition: namely, libertarian socialism (or anarchism). Day's "politics of affinity" and his advocacy of "structural renewal" simply exemplify the politics of "old" anarchism—the anarchism of Kropotkin and a host of social or class struggle anarchists from Goldman, Rocker, and Landauer in the past to Bookchin and the Anarchist Federation in the present. Thus there is very little that is new or original in the so-called "postanarchism."

4. Conclusion

We can but conclude that there is nothing particularly "new" about the "new anarchism" and the anarchistic tendencies that are being expressed in the new social movements or in the anti-globalisation protests is not some "new mode" of anarchism or some "new paradigm" but the kind of social or class struggle anarchism that had its origins long ago in working class struggles for libertarian socialism. This form of anarchism, as earlier indicated, is still the most vibrant form of anarchism around (see Sheehan 2003, Franks 2006).

Postscript

This paper is the gist of a talk I gave to the Northern Anarchist Network in September 2007. It aims to be a polemical and radical critique of so-called new anarchism. It is not intended to be an ego trip on my part, or intentionally sneering or dismissive of other radical scholars, and generalisations are unavoidable in discussing such vague concepts as "new anarchism" and "postmodernism." Unsurprisingly, the paper was rejected by *Anarchist Studies* as too polemical and "unscholarly" for academic sensibilities.

References

Albert, Michael. 2006. *Realizing Hope: Life beyond Capitalism*. London: Zed Books.
Baldwin, Roger N. 1927. *Kropotkin's Revolutionary Pamphlets*. New York: Dover.
Bey, Hakim. 1991. *T.A.Z.: The Temporary Autonomous Zone, Ontological Anarchy, Poetic Terrorism*. Brooklyn: Autonomedia.
_____. 1994. *Immediatism*. San Francisco: AK Press.
Black, Bob. 1997. *Anarchy after Leftism*. Columbia, MO: CAL Press.
Bookchin, Murray. 1980. *Toward an Ecological Society*. Montreal: Black Rose Books.
_____. 1982. *The Ecology of Freedom*. Palo Alto: Cheshire.
_____ et al. 1993. *Deep Ecology and Anarchism*. London: Freedom Press.
_____. 1995. *Re-enchanting Humanity*. London: Cassell.
_____. 1995b *Social Anarchism or Lifestyle Anarchism*. San Francisco: AK Press.
_____. 1999. *Anarchism, Marxism, and the Future of the Left*. San Francisco: AK Press.

Bourdieu, Pierre. 1998. *Acts of Resistance: Against the Tyranny of the Market*. Cambridge: Polity Press.

Bunge, Mario. 1996. *Finding Philosophy in Social Science*. New Haven: Yale University Press.

Callinicos, Alex. 1997. "Postmodernism: A Critical Diagnosis." In *Great Ideas Today*, 206–55. Chicago: Encyclopedia Britannica.

Clark, John P. 1976. *Max Stirner's Egoism*. London: Freedom Press.

Clark, John P., and Camille Martin, eds. 2013. *Anarchy, Geography, Modernity: The Radical Social Thought of Elisée Reclus*. Oakland: PM Press.

Curran, Giorel. 2006. *21st Century Dissent: Anarchism, Anti-Globalization and Environmentalism*. Basingstoke: Palgrave.

Day, Richard J.F. 2005. *Gramsci Is Dead: Anarchist Currents in the Newest Social Movements*. London: Pluto Press.

Derrida, Jacques. 1976. *Of Grammatology*. Baltimore: John Hopkins University Press.

Detmer, David. 2003. *Challenging Postmodernism: Philosophy and the Politics of Truth*. Amherst: Humanity Books.

Douglas, Mary. 1975. *Implicit Meanings: Essays in Anthropology*. London: Routledge and Kegan Paul.

Ellen, Roy F. 1986. "What Black Elk Left Unsaid." *Anthropology Today* 2, no. 16: 8–12.

Flax, Jane. 1990. *Thinking Fragments: Psychoanalysis, Feminism, and Postmodernism in the Contemporary West*. Berkeley: University of California Press.

Foucault, Michel. 1977. *Language, Counter-Memory, Practice*. Ithaca: Cornell University Press.

Franks, Benjamin. 2006. *Rebel Alliances: The Means and Ends of Contemporary British Anarchisms*. Oakland: AK Press.

Gellner, Ernest. 1992. *Postmodernism, Reason and Religion*. London: Routledge.

Goaman, Karen. 2004. "The Anarchist Travelling Circus." In *Changing Anarchism: Anarchist Theory and Practice in a Global Age*, edited by Jon Purkis and James Bowen, 163–80. Manchester: Manchester University Press.

Hay, Peter. 2002. *A Companion to Environmental Thought*. Edinburgh: Edinburgh University Press.

Heidegger, Martin. 1994. *Basic Questions of Philosophy*. Bloomington: Indiana University Press.

Holloway, John. 2005. *Change the World without Taking Power*. London: Pluto Press.

Kinna, Ruth. 2005. *Anarchism: A Beginner's Guide*. Oxford: Oneworld Publications.

Kropotkin, Peter. 1902 *Mutual Aid: A Factor in Evolution*. London: Heinemann.

———. 1970. *Selected Writings on Anarchism and Revolution*. Edited by Martin A. Miller. Cambridge, MA: MIT Press.

Leier, Mark. 2006. *Bakunin: The Creative Passion*. New York: St. Martin's Press.

López, José, and Garry Potter, eds. 2001. *After Postmodernism: An Introduction to Critical Realism*. London: Athlone.

Lyotard, Jean-François. 1984. *The Postmodern Condition: A Report on Knowledge*. Manchester: Manchester University Press.

Marx, Karl, and Friedrich Engels. 1965. *The German Ideology*. London: Lawrence and Wishart.

McLaughlin, Paul. 2002. *Mikhail Bakunin: The Philosophical Basis of His Anarchism*. New York: Algora.

Monbiot, George. 2000. "No Way to Run a Revolution." *The Guardian*, May 10, 2000.

Moore, John. 2004. "Lived Poetry: Stirner, Anarchy, Subjectivity, and the Art of Living." In *Changing Anarchism: Anarchist Theory and Practice in a Global Age*, edited by Jon Purkis and James Bowen, 55–72. Manchester: Manchester University Press.

———, ed. 2004b. *I Am Not a Man, I Am Dynamite: Friedrich Nietzsche and the Anarchist Tradition*. Brooklyn: Autonomedia.

_____. 2004c "Attentat Art: Anarchism and Nietzsche's Aesthetics." In *I Am Not a Man, I Am Dynamite: Friedrich Nietzsche and the Anarchist Tradition*, edited by John Moore, 127–42. Brooklyn: Autonomedia.

Morland, David. 2004. "Anti-Capitalism and Poststructuralist Anarchism." In *Changing Anarchism: Anarchist Theory and Practice in a Global Age*, edited by Jon Purkis and James Bowen, 23–38. Manchester: Manchester University Press.

Morris, Brian. 1982. *Forest Traders* London: Athlone Press.

_____. 1986. "Deforestation in India and the Fate of the Forest Tribes." *The Ecologist* 16: 253–57 in *Ecology and Anarchism*, 72–78. Malvern Wells: Images.

_____. 1993. *Bakunin: The Philosophy of Freedom*. Montreal: Black Rose Books.

_____. 1996. *Ecology and Anarchism: Essays and Reviews on Contemporary Thought*. Malvern Wells: Images.

_____. 1997. "In Defence of Realism and Truth: Reflections on the Anthropological Followers of Heidegger." *Critique of Anthropology* 17, no. 3: 313–40.

_____. 2004. *Kropotkin: The Politics of the Community*. Amherst, NY: Humanity Books.

_____. 2005. *Anthropology and Anarchism: Their Effective Affinity*. Goldsmiths College: Anthropology Dept. Research Papers No.11

_____. 2006. *Religion and Anthropology: A Critical Introduction*. Cambridge University Press.

_____. 2007. "Nietzsche and Anarchism." Review of John Moore, ed., *I Am Not a Man, I Am Dynamite* (2004). *Social Anarchism* 40: 54–57

Newman, Saul. 2002. "Max Stirner and the Politics of Humanism" *Contemporary Political Theory* 1: 221–38.

_____. 2004. "Anarchism and the Politics of Ressentiment." In *I Am Not a Man, I Am Dynamite: Friedrich Nietzsche and the Anarchist Tradition*, edited by John Moore, 107–26. Brooklyn: Autonomedia.

Purkis, Jon, and James Bowen, eds. 1997. *Twenty-First Century Anarchism: Unorthodox Ideas for the New Millennium*. London: Cassell.

_____. 2004. *Changing Anarchism: Anarchist Theory and Practice in a Global Age*. Manchester: Manchester University Press.

Rorty, Richard. 1979. *Philosophy and the Mirror of Nature*. Princeton, NJ: Princeton University Press.

Rötzer, Florian. 1995. *Conversations with French Philosophers*. Atlantic Highlands, NJ: Humanities Press.

Sheehan, Sean M. 2003. *Anarchism*. London: Reaktion Books.

Sheppard, Brian Oliver. 2003. *Anarchism vs. Primitivism*. Tucson, AZ: See Sharp Press.

Stirner, Max. 1973. *The Ego and His Own*. New York: Dover Publications.

Tormey, Simon. 2004. *Anti-Capitalism: A Beginner's Guide*. Oxford: Oneworld Publications.

Tyler, Stephen. 1986. "Postmodern Ethnography." In *Writing Culture: Writing Culture: The Poetics and Politics of Ethnography*, edited by James Clifford and George Marcus, 122–40. Berkeley: University of California Press.

Ward, Colin. 1973. *Anarchy in Action*. London: Allen and Unwin.

Wolin, Richard. 2004. *The Seduction of Unreason*. Princeton, NJ: Princeton University Press.

Wood, Ellen Meiksins, and John Bellamy Foster, eds. 1997. *In Defense of History: Marxism and the Postmodern Agenda*. New York: Monthly Review Press.

Zerzan, John. 1998. *Elements of Refusal*. Seattle: Left Bank Books.

_____. 1994. *Future Primitive and other Essays*. Brooklyn: Autonomedia.

_____. 2002. *Running on Emptiness*. Los Angeles: Feral House.

10

Rudolf Rocker: The Gentle Anarchist (2009)

Prologue

In this essay I want to pay homage to the memory and teachings of the anarcho-syndicalist Rudolf Rocker, who died fifty years ago.

These days Rocker is very much a forgotten figure, even within anarchist circles. There is no mention of Rocker, for instance, in Daniel Guérin's anthology of anarchism, *No Gods, No Masters* (1998), nor in George Woodcock's *The Anarchist Reader* (1997)—and he is singularly absent from Bookchin's magnum opus *The Ecology of Freedom* (1982). And even in general accounts of the history of the anarchist movement, as his biographer, Mina Graur (1997) recounts, Rocker is mentioned only in passing.

With the emergence, according to many academics and literary scholars, of the so-called "new" anarchism, Rocker is in the process of being further marginalised. For the anarchism of an earlier generation of libertarian socialists is declared to be "old" anarchism, and is thus perceived to be old-fashioned and outdated. "Old" anarchism, we are informed, has become a "historical baggage" that needs to be rejected, or at least given a "major overhaul" (Purkis and Bowen 1997, 3); it is a historical relic of no relevance at all to contemporary radical activists (Holloway 2005, 21).

The notable Ghanaian writer Ayi Kwei Armah (1979) remarked that "the present is where we get lost—if we forget our past and have no vision of the future." There is then a growing tendency among many radical scholars not only to forget the past but to repudiate it entirely, thus denying that an earlier generation of anarchists have anything to teach us with regard to contemporary struggles. Such myopia seems to me unwarranted and unfair to a past generation of anarchists. So my aim here is to briefly outline some of the teachings and basic ideas of Rudolf Rocker, thus indicating their contemporary relevance. For it seems to me that we should not look upon Rocker simply as a historical relic, of interest only to historians, but having a contemporary relevance and as a source of inspiration to all those today who strive for radical change, whatever the kind of anarchism.

Anarchism is both a social movement and a political tradition. It is a movement that did not simply wither away at the end of the 1930s (as Mina Graur contends [1997, 244]) only to re-emerge phoenix-like as a "new" form of anarchism in the anti-globalisation demonstrations in Seattle in 1999 (as David Graeber [2002] contends). For class struggle anarchism has had a decided and continuing presence in all the protest movements since the Second World War.

Before outlining Rocker's anarchism, let me briefly sketch his biography.

Rocker: A Biographical Sketch

Rudolf Rocker was born in Mainz on the Rhine, South Germany, in 1873. He lost both his parents early in life, and was orphaned at the age of fourteen. He was sent to a Catholic orphanage, an experience he likened to a desert exile. Apprenticed to a bookbinder, he was influenced by his maternal uncle Rudolf Naumann, who introduced him to socialism as well as to books, especially romantic literature. Throughout his life Rocker was an incurable romantic, and as his son Fermin records, he had an absolute passion and reverence for books (1998, 23). Like myself, Rocker seems to have spent many hours scouring second-hand bookshops for cheap books.

At the age of seventeen, Rocker joined the German Social Democratic Party but soon became critical of its rigid and authoritarian ethos, and its statist politics, that derived essentially from the Marxists Friedrich Engels and Ferdinand Lassalle. Expelled from the Party, Rocker began associating with anarchists, and studying the writings of Bakunin, Kropotkin, and Johann Most. Thus by the age of twenty Rocker had become an anarchist. To avoid arrest Rocker left Germany in November 1892, and after a two-year sojourn in Paris, came to London at the end of 1894. Rudolf Rocker's subsequent life as an anarchist consists of three distinct phases.

For some twenty years (1895–1914) Rocker was an "anarchist missionary to the Jews," as William Fishman (1975) describes Rocker in his excellent study of East End Jewish radicals in the years before the First World War. For as Rocker himself graphically describes in his autobiography (2005), Rocker became closely identified with the Jewish immigrant workers of the London East End. He married a young Ukrainian Jew, Milly Witcop, who was to become a lifelong partner, lived in Stepney Green, and learned to speak and write the Yiddish language. He edited the Yiddish anarchist newspaper *Arbeter Fraynd* (Worker's Friend) and a literary monthly journal, *Germinal*, and was actively involved in the Jewish labour unions in their fight for better working conditions.

Two points are worth noting about Rocker's "London years" which he described incidentally as the best years of his life.

The first is that not only was Rocker deeply involved in political struggles and spreading the anarchism message through propaganda, but he regularly gave lectures on literary and cultural topics at the Jubilee Street Club, which was then the centre for Jewish Social and intellectual life in the East End. By all accounts Rocker was an excellent and inspiring lecturer, and continued to give lectures to diverse audiences throughout his long life. Nellie Dick recalls that Rocker's lectures always held his audience "spellbound" (Avrich 2005, 283).

Secondly, Rocker came to know personally, and to pen memorable sketches of the many anarchists who were involved in radical struggles in the years prior to the First World War—for example, Louise Michel, Errico Malatesta, John Turner, Max Nettlau, Gustav Landauer, Emma Goldman, Sam Yanovsky, and Peter Kropotkin.

Unlike his brother-in-law Guy Aldred, who was married to his wife's younger sister Rose Witcop, Rocker had a very high regard for Peter Kropotkin, both as a scholar, and as an anarchist, describing him as a man "of great personal charm and kindliness, with all his great learning modest and unassuming and with a burning passion for justice and freedom" (2005, 75). But with the outbreak of the First World War, Rocker, along with Malatesta, was highly critical of Kropotkin's pro-war stance, but nevertheless they always remained close friends.

But it is of interest that Rocker makes no mention in his memoirs of Guy Aldred, for he appears to have considered Aldred rather brash and his anarchism rather shallow (Fermin Rocker 1998, 66).

Given his German background and political radicalism, Rocker was arrested at the outbreak of the First World War, and interned as an "enemy alien." At the end of the war he was deported from Britain to the Netherlands, but soon made his way to Germany. There began the second phase of his life (1918–1933), and it was a period that the anarcho-syndicalist Sam Dolgoff considered to be the most crucial of Rocker's career—constituting his "greatest achievement" (1986, 109). For Rocker was an eloquent public speaker, and travelled throughout Germany giving propaganda speeches promoting libertarian socialism. He was actively involved and one of the founders of the Free German Workers Union, and according to Nicolas Walter was the "moving spirit" of the International Congress held in Berlin in 1922, which led to the formation of the International Working Men's Association. Along with Alexander Shapiro and Agustín Souchy, he was one of its secretaries. Indeed Rocker became one of the leading advocates of anarcho-syndicalism, and used his influence to counter both parliamentary socialism and Bolshevism. Rocker was also critical of the "organisational platform" drafted by the Russian anarchist exiles in Paris (1926), the Dielo Trouda (Workers' Cause) group. Nestor Makhno and Peter Arshinov were its main authors, and it emphasised the need for anarchist organisation, akin almost to a political party.

During this phase of his life, Rocker made several extensive lecture tours to the United States, developed close contacts with leading figures in the Russian anarchist movement, such as Voline (Vsevolod Eichenbaum) and Gregory Maximoff; wrote an important biography of Johann Most (1924); and even debated with the Nazi Otto Strasser on the issue of race and nationalism, for Rocker led the libertarian opposition to the rising tide of National Socialism (Graur 1997, 168–74). But when Hitler came to power in 1933, the burning of the Reichstag and the murder of his friend Erich Mühsam meant that Rocker and his wife had to flee Germany. He left behind all his belongings, including a library of around five thousand books (which were probably burned by the Nazis) and took with him only the manuscript *Nationalism and Culture*, which he had been writing for more than a decade and had then just completed. Thus began the last phase of Rocker's life, as an exile in the United States (1933–1958).

In his "new exile," as Rocker described it, he began writing his memoirs, urged on by his friend Max Nettlau. After spending some years at Towanda, a small village in Pennsylvania, where Milly's sister Fanny lived, Rocker eventually settled in 1937 at the Mohegan colony, some forty miles from New York. That same year he published his magnum opus *Nationalism and Culture*, supported by his friends. A scholarly analysis describing the rise of the modern state and the ideology of nationalism, the book was widely praised by his contemporaries. Albert Einstein considered the book "extraordinarily original and illuminating" while Bertrand Russell thought it penetrating and highly informative and an "important contribution to political philosophy." The well-known philosopher Will Durant regarded it as a "magnificent book written with profound understanding of man and history" (Graur 1997, 177).

Interestingly, *Nationalism and Culture* has been singularly ignored by academics, and none of the well-known theorists of nationalism (e.g., Anderson 1983, Gellner 1983, Hobsbawm 1990) ever mention Rocker's seminal work.

At the time of the Spanish Civil War, Rocker wrote two important and influential pamphlets, entitled "The Truth about Spain" (1936) and "The Tragedy of Spain" (1937) and, urged by Emma Goldman, wrote an excellent, short introduction to *Anarcho-Syndicalism* (1938). But just like Kropotkin in the First World War Rocker sided with the Allies at the outbreak of the Second World War, believing that a German victory and the triumph of fascism would be an absolute disaster for humanity. Although supported by other anarchists, such as Gregory Maximoff and Sam Dolgoff, most anarchists felt that Rocker's "pro-war" stand was contrary to anarchist principles. Rocker was sharply criticised by Marcus Graham, as well as by the Freedom Group. Indeed Vernon Richards and the Freedom Group were consistently hostile towards Rocker, suggesting that he was patronising towards Jewish workers, that his influence within the anarchist movement was highly exaggerated, and that Rocker was the focus of a personality cult, as well as being surrounded by wealthy sycophants. Yet, as Dolgoff stressed, Rocker was a very humble and modest person, and never sought any kind of hero-worship let alone encouraged it (Avrich 2005, 423).

Even in his old age, in spite of bitter disappointments, and though abandoning many aspects of his earlier revolutionary anarchism, Rocker, like Agustín Souchy, remained faithful, Sam Dolgoff suggests, to the anarchist ideal (1986, 119). Souchy, in fact, considered Rocker as the "spiritual" leader of the German anarchists during the inter-war period, and described Rocker as honest, good-natured, and tolerant, a brilliant speaker, and an incisive social theorist who had an extraordinary knowledge of the history of anarchism and the international labour movement (Souchy 1992, 49).

Reading the many recollections of Rocker in Paul Avrich's *Anarchist Voices*, one is clearly given the impression that Rocker was indeed a "gentle" anarchist, a warm, kind, engaging person, who "exuded humanity" and yet was extremely learned (Avrich 2005, 253).

Let me now turn to an exposition of Rocker's anarchist philosophy.

The Political Legacy of Rocker
1. Anarchism: Libertarian Socialism
It is important to recognise that Rocker always explicitly identified himself as an anarchist, as a libertarian socialist, and in his autobiography he continually affirms his identity as an anarchist. As he wrote: "I am an anarchist not because I believe anarchism is the final goal, but because I believe there is no such thing as a final goal. Freedom will lead us to continually wider and expanding understanding and to new forms of social life" (2005, 111). Chaining ourselves to programmes, dogmas, past ideas, always leads, Rocker felt, to tyranny.

Like Kropotkin, Rocker recognised that anarchist ideas had existed throughout human history, and were expressed in the writings of Lao Tzu and the Stoics, as well as in the peasant revolts in Europe from the thirteenth to sixteenth century. But modern anarchism, as a political movement and tradition, essentially emerged in the nineteenth century. It was, he felt, the "confluence" of two great currents of thought that had their origins in the radical aspects of the Enlightenment and the French Revolution—liberalism and socialism (1938, 21).

Liberalism, as expressed by Jefferson, Diderot, and Wilhelm Von Humboldt, put an emphasis on the freedom of the individual over his or her own person. Von Humboldt, in particular, strongly appealed to Rocker—as he does to Chomsky; for Von Humboldt stressed that the main purpose of human life was that every human being should develop his or her own powers and personality. Thus freedom of the individual was a necessary condition for the development of human culture. Classical liberalism therefore expressed the idea of limiting the functions of government and state power to a minimum. Anarchism shares with liberalism the idea that the freedom and happiness of the individual is paramount, but takes Jefferson's idea that the government is best that governs least a step further, and declares, with Thoreau, "that government is best which governs not at all" (Rocker 1938, 23).

Socialism, on the other hand, involved the idea that social justice and equality were only possible if economic monopolies were eliminated. It entailed the idea that land and the means of production should be under social ownership, and that the production of the basic necessities of human life and the distribution of social wealth should be undertaken by cooperative labour and voluntary associations. Rocker recognised, of course, that socialism had been embraced by people of every political persuasion, from theocracy to caesarism, including the National Socialism of the German fascists, the welfare state of the democratic socialists, as well as the state capitalism of the Bolshevik tyranny. But for Rocker, as for Bakunin and Kropotkin, socialism would be free, and based on voluntary principles, or it would not be true socialism. Thus, for Rocker, socialism meant a society based on voluntary associations and mutual aid and he thus repudiated private property, the market economy, and wage labour.

Rocker therefore considered anarchism to be a synonym of libertarian socialism, entailing the repudiation of both capitalism (and all the forms of economic exploitation) and the state (and all political and coercive institutions within society) (1938, 9). He thus strongly affirmed Bakunin's famous adage: "That liberty without socialism is privilege and injustice, and that socialism without liberty is slavery and brutality" (Lehning 1973, 110).

Thus while Rocker emphasised that anarchism was a form of socialism, his experience with the German Social Democrats (Marxists) led him to fear that socialism without liberty would inevitably lead to tyranny (2005, 32). Rocker was thus an advocate of what he described as "free socialism" (103).

On the other hand, Rocker was opposed to extreme egoism or to abstract conceptions of freedom, and held that true freedom existed only when it was fostered by a spirit of personal responsibility and a sense of social solidarity. He held that only when an ethical feeling of responsibility towards one's fellow humans is combined with an urge for social justice is real freedom possible. Without socialism and social solidarity, individual freedom, Rocker argued, leads only to "unlimited despotism and the oppression of the weak by the strong" (1937, 96).

Thus Rocker felt that one cannot be free, either politically or personally, so long as a person is in economic servitude of another, and therefore equality is a necessary condition for human freedom. But Rocker emphasised that freedom is not an abstract philosophical concept but a real concrete possibility, enabling people to develop their own inherent powers, capacities, and talents (1937, 167; 1938, 31).

As Rocker wrote: "Equality of economic conditions for each and all, is always a necessary precondition for human freedom, but it is never a substitute for it" (1937, 237). Authoritarian socialism, he felt, was actually a contradiction in terms: socialism would be free and voluntary or it would not be socialism. Anarchism, Rocker therefore concluded, was a synthesis of liberalism and socialism; and he agreed with both Proudhon and Bakunin that "socialism" without freedom was the worst form of slavery (1938, 28).

2. Capitalism and State Power: Their Symbiotic Relationship

Again like Kropotkin, Rocker always stressed that there was a close and intrinsic relationship between capitalism (economic exploitation) and state power (political oppression). As he put it: "The exploitation of man by man and the domination of man over man are inseparable, and each is the condition of the other" (1938, 28).

Thus economic exploitation and political oppression, Rocker felt, mutually support each other, the function of the state being essentially to defend social privileges and class exploitation. Anyone who fails to recognise this, he argues, does not understand the real nature of the present social order (1938, 30).

In his discussion of the rise of the modern state during the Renaissance period, Rocker thus argues that the emerging mercantile corporations needed a strong political power, with necessary military force, to recognise and protect their interests, and that it was from these small city-states, associated with Venice and Genoa, that the modern nation-state emerged. This was to have a revolutionary impact on all European thought and institutions (1937, 94).

Thus Rocker concluded that the will to power always leads to the exploitation of the weak, and that "every form of exploitation finds its visible expression in a political structure" which functions to uphold the system of exploitation (1937, 93).

3. Critique of Marxism

Rocker was a lifelong critic of Marxism, both on epistemological and political grounds.

Although Rocker recognised the existence of cosmic laws, and acknowledged the importance of causal factors in the understanding of human social life, he always made a clear distinction between the laws relating to physical or natural events, and the causes that underlie the processes of social life. He emphasised that social institutions are human creations, and so cannot be understood in terms of fixed, deterministic processes. This led him to repudiate Marx's economic theory of history, which he felt was consonant with Marx's rather deterministic theory, a theory which suggested that economic factors were the primary forces in shaping human history.

Rocker never denied the importance of economic factors; he only suggested that there were many events in history—the Crusades, the Reformation, Inca power, for example—that could not be explained by purely economic factors. As he wrote: "There is scarcely an historical event to whose shaping economic causes have not contributed, but economic forces are not the only motive powers which have set everything else in motion" (1937, 28).

In understanding the evolution of social life, many factors, he argued, therefore have to be taken into account, especially the "will-to-power," and the economic factor should neither be overlooked nor unduly exaggerated (1937, 32).

Rocker always paid a warm tribute to the writings of Pierre-Joseph Proudhon and Michael Bakunin—in critiquing the Jacobin tradition of the early socialists (Babeuf, Louis Blanc), and in laying the foundations of modern anarchism—the libertarian tradition that repudiated both economic monopolies and the modern nation-state. Such a tradition advocated the need to reconstruct social life on a federalist basis, in the form of free and autonomous communities focused around workers' control and voluntary associations.

Rocker, therefore, followed Proudhon and Bakunin in being highly critical of Marx's brand of socialism. Although he acknowledged Marx's outstanding intellect and enormous erudition, Rocker nevertheless felt that Marx had not enriched socialism with any creative thought. Marx was essentially a brilliant analyst of the capitalist system. And quoting a letter Marx wrote to his friend Friedrich Engels (July 1870), Rocker highlights the fact that Marx had a very authoritarian bent, celebrating the centralisation of state power, and the centralisation of the German working class

(1937, 234). Marx, Rocker argued, along with his followers Engels, Lassalle and Lenin, were all steeped in the ideas of Hegel, and strong advocates of achieving socialism through the conquest of political power, whether by parliamentary means or through a vanguard party. Rocker, from his earliest years, always feared that socialism without liberty would inevitably lead to tyranny. Democratic socialism and the Bolshevik experience in Russia, Rocker argued, had shown that the union of state power with socialism inevitably leads not to any form of socialism but to state capitalism and tyranny.

4. Religion and Power

Rocker always made a clear distinction between three essential concepts, namely, religion, power, and culture, although he recognised that all three aspects of human life have their roots in social existence and the human instinct for self-preservation.

Religion he defined as a "feeling of human dependence" on some "higher power" which the human imagination had brought into being (1937, 45). Religion therefore implied, of its very nature, a hierarchical relationship, and a mode of dominion, and he remarked that the concept of God was the epitome of all power. This led Rocker to suggest that all politics is, in the last instance, religion; that all power is made in the image of god. As he put it: "All power has its roots in God: all rulership is in its inmost essence divine" (1937, 48).

Suggesting that the earliest forms of power had their origins in conquest, Rocker stressed that no power can in the long term rely on brute force alone. He therefore argued that there has always been a close and intricate relationship between political power and religion, whatever form these may take. He outlines their close relationship in all the early empires—India, Japan, China, Inca, as well as that of Genghis Khan and the Absolutist States of Europe (1937, 49–54). Every system of government, Rocker concluded, has a certain theocratic character (1937, 55), and he held that no state or system of power can maintain itself for long without basing itself on some form of religious consciousness, thus, no temporal power has ever been able to dispense with religion, Rocker contended, and even the Bolshevik state turned atheism into a religion and Lenin into a religious icon. And the cult of the nation, he argued, was in fact a new religion, the ideology of the modern nation-state (see below).

But though stressing the inherent political nature of religion, this did not imply for Rocker either an intolerant attitude towards religion, or a disavowal of the essentially spiritual nature of human life. As he wrote: "People must learn tolerance. It would never occur to me to upset anyone engaged in his religious devotions" (2005, 81).

The right to act according to one's religious beliefs belongs to everyone, and Rocker always stressed the importance of having a diversity of views and opinions. The important point is that people must think and act for themselves.

The spiritual nature of humans—the human spirit—was expressed, Rocker argued, not in religious beliefs and rituals, but in various modes of thought and activity—"in science, art and literature, in every branch of philosophical thought and aesthetic feeling"—which he felt must always be the common cultural heritage of all humans (2005, 84).

5. On Nationalism

Rocker's key work *Nationalism and Culture* (1937) is essentially concerned with exploring the rise of the nation-state and the ideology of nationalism within the European context.

The Renaissance was the key period for Rocker, for it not only involved the disintegration of local communities and the rise of the modern state, but also the rise of individualism and the emergence of nationalism as a surrogate religion. A key figure here, for Rocker, is Machiavelli. Thus from around the sixteenth century, with the decline in the political powers of the Catholic Church and that of the absolutist states, there developed what Rocker describes as the nationalist state (1937, 16), state systems that focused around a specific territory and the ideology of the "nation."

This conception of the state was the doctrine of democracy, which Rocker viewed as the complete antithesis of liberalism. He specifically relates the democratic ideal to Rousseau, whom he describes as the "state theoretician" par excellence, for Rousseau introduced the Jacobin idea of freedom, with its emphasis on the general will, law, and state authority (1937, 167–70). Thus by the end of the eighteenth century, Rocker argues, a new faith had emerged, that of nationalism—"the religion of the democratic state" (1937, 179).

Rocker makes a distinction between a "people," always a local community with rather narrow boundaries, which had existed long before the emergence of the state, and the "nation," which for Rocker was essentially a political notion.

For Rocker, the "nation" was not so much an "imagined community," as Benedict Anderson (1983) contended, as a cultural artefact that was intrinsically wedded to that of the modern state. There is no close identity between nation and language, Rocker argued, nor is it possible to delineate a clearly defined culture, though there is a clear distinction between one's natural attachment to a home—manifested in the human enjoyment of nature— and patriotism, and the love of a nation-state (1937, 24). Thus Rocker concluded that nationalism has never been anything but the "political religion" of the modern state (1937, 201); that it is reactionary in its very nature (213); and that nationalism reaches its apotheosis in the fascist state.

Rocker tended to affirm the close relationship between democracy and the state, particularly with regard to Jacobin politics, and to suggest that liberalism is antithetical both to democracy and to the religion of nationalism. He quoted from the radical liberal Tom Paine: "The world is my country, all men are my brothers" (1937, 203).

Josep Lobera significantly titled his study of nationalism in Western Europe *The God of Modernity* (1994). This, of course, was Rocker's own thesis, namely nationalism is a religious conception, intrinsically linked to the rise of the modern state.

But one problem with Rocker's analysis is that, unlike Kropotkin, he lacked any anthropological perspective, and thus never theorised exactly what constituted a "people" or "homeland," or if a distinction could be made between ethnicity and nationalism.

6. Power and Culture

Although Rocker recognised that both power and culture had a common source, in human social activities, he nevertheless felt that in historical terms that there had always been a tension or opposition between the institutions of political and economic power, and the cultural activities of people. Power, for Rocker, always pre-supposed some form of equality or slavery, and always implied a negative element—coercion, control, law, command, uniformity.

In contrast, culture for Rocker was a creative force, arising spontaneously from cooperative social activities. It was not something restricted to a particular ethnic or national context. Rocker likened culture to a tropical Banyan tree that was continually spreading its branches, putting down roots as they touched the earth (1937, 83). Culture expressed the "spiritual life" of humans as species-being—as manifested in science, philosophy, literature, and the arts, and in human aspirations for equality and freedom.

There are no pure or isolated cultures, Rocker argued, for culture is something common to all humans—for "we are all children of this earth and subject to the same laws of life" (1937, 436).

Thus Rocker came to see an inverse relationship between culture and power. In contexts of political disunity, such as in classical Greece or in the Arabic culture of Spain, there was, he argued, a flourishing of human culture, while periods of political centralisation, as in classical Rome, only tended to stifle cultural creativity.

Rocker emphasised that although political rights and liberties may be encapsulated in state constitutions—the right to assembly, to form trade unions, freedom of the press, et al.—these have always been obtained by the struggles of working people. And that governments "are ever ready to curtail existing rights or to abolish then entirely" if they feel there will be no resistance from the people (1937, 88).

7. Anarcho-Syndicalism

Rocker was both a leading activist in the anarcho-syndicalist movement, as well as one of its foremost theorists. His *Anarcho-Syndicalism* (1938) is a classic of anarchist literature, and continues to be reissued almost every decade.

Anarcho-syndicalism is a form of libertarian socialism—a mode of action—that attempts to combine an emphasis on the class struggle, and the emancipation of the working class, with anarchist principles, in being both anti-state and anti-capitalist.

Anarcho-syndicalism as it developed in the early years of the twentieth century, had the following characteristics: Firstly, it repudiated entirely the anarchist tactic of "propaganda by the deed," direct acts of violence against the bourgeois state by assassination or terrorism. Secondly, it critiqued and rejected the parliamentary road to socialism, as advocated by the British Labour Party and the German Social Democrats (Lassalle) and various other socialist parties. And finally, anarcho-syndicalists, like the early followers of Bakunin, repudiated entirely the Marxist concept of the "dictatorship of the proletariat" by means of a revolutionary vanguard party.

Rocker held that the parliamentary road to socialism was simply reformist, and led to the incorporation of socialist ideas into the bourgeois state, while the seizure of power by a revolutionary minority on behalf of the working class, as with the Bolsheviks, would lead only to bureaucratic state capitalism (1938, 85).

Anarcho-syndicalists put a fundamental emphasis on the class struggle, and on the role of the trade unions, which were viewed as having a dual purpose: As Rocker put it: "The trade union, the syndicate, is the unified organisation of labour and has for its purpose the defence of the interests of the producers within existing society and the preparing for and practical carrying out of the reconstruction of social life after the pattern of socialism" (1938, 86).

Anarcho-syndicalism was thus based on direct action, solidarity, workers' self-management and federalist principles. It was not against political struggle as such, for as Rocker put it: every event that affects the life of a community is of a political nature (1938, 115). It was rather against the idea that anything radical could be achieved through participation in the parliamentary system, or by a vanguard party appropriating state power. Rocker bewailed the fact that the Bolsheviks, in their fanatical zeal for government, had betrayed the socialist revolution in Russia (1938, 95).

How have people responded to Rocker's anarchism and his espousal of anarcho-syndicalism? Essentially people have responded in three ways:

Firstly, advocates of the so-called "new" anarchism have tended to express a rather dismissive attitude towards an earlier generation of anarchists, including Rocker. The "old" anarchism has been dismissed as "obsolete" or "outmoded." A typical example is Richard Griffin's essay entitled "Reality Check" (2002). Not only does Griffin deny the reality of capitalism and the state (suggesting that one can simply imagine them away!) but he suggests that the postmodernists are the real ultra-radicals, who have repudiated revolution and have ceased to be anti-capitalist (the idea of class struggle is outmoded), and have instead become involved in creating a "truly new social world." Rocker, we are informed, was not about creating a new world but only in improving the "existing one" (capitalism!) and Rocker is alleged to have poured scorn on libertarian education, communes and voluntary associations. Griffin concludes that we need to bring about a world based on mutual aid and cooperation—as if he was saying something new and original!

Rocker, of course, like Kropotkin and other class struggle anarchists, did not see any contradiction between class struggle—opposing capitalism and all forms of hierarchy and power—and advocating mutual aid, autonomous communities, voluntary associations and workers' self-management. Griffin is simply ignorant of what Rocker's anarchism involved.

The second response to Rocker's anarcho-syndicalism is that expressed by Kropotkin and Malatesta long ago, and Murray Bookchin more recently. This suggests that although it is necessary to participate in the working class movement and to support the workers' struggle against capitalism and the state, to focus primarily on the factory system and on the general strike is far too narrow and limiting. For it ignores the crucial importance of local cooperatives, agrarian communes and the city as a bioregional community. It is however fair to say that Rocker was as much influenced by Kropotkin's communalism as he was by Bakunin's anarcho-syndicalism, and like a later generation of anarcho-syndicalists—Maximoff, Dolgoff, Meltzer—he had a fairly broad conception of anarchism. Rocker certainly did not believe that the general strike would bring the capitalist system down overnight (1938, 121) and acknowledged the importance of the Spanish revolution as an exemplification of libertarian socialism in action (1938, 98).

In his later years Rocker published a study of American anarchism, *Pioneers of American Freedom* (1949) and tended to replace anarcho-syndicalism with "community socialism," the community replacing the trade unions as the primary unit of society (Graur 1997, 237). What Rocker did lack, however—and in this he contrasted markedly with Kropotkin—was any ecological perspective.

Finally, it has to be noted that many of Rocker's ideas still find a resonance in many of the principles expounded by contemporary anarcho-syndicalists, as reflected for instance in such radical magazines as *Black Flag* (London), *The Rebel Worker* (Sydney), and *Anarcho-Syndicalist Review* (Philadelphia).

We can but conclude with a quotation from Rocker, the gentle anarchist: "My innermost conviction was that anarchism was not to be conceived as a definite closed system nor as a future millennium but only as a particular trend in the historic development towards freedom in all fields of human thought and action" (2005, 73).

References

Anderson, Benedict. 1983. *Imagined Communities: Reflections on the Origin and Spread of Nationalism.* London: Verso.

Armah, Ayi Kwei. 1979. *The Healers: An Historical Novel.* Popenguire: Senegal: Per Ankh; London: Heinemann.

Avrich, Paul. 2005. *Anarchist Voices: An Oral History of Anarchism in America.* Oakland: AK Press.

Bookchin, Murray. 1982. *The Ecology of Freedom.* Palo Alto: Cheshire.

Dolgoff, Sam. 1986. *Fragments: A Memoir.* Cambridge: Refract Press.

Fishman, William J. 1975. *East End Jewish Radicals, 1875–1914.* London: Duckworth.

Gellner, Ernest. 1983. *Nations and Nationalism* Oxford: Blackwell.

Graeber, David. 2002. "For a New Anarchism." *New Left Review* 13: 61–73.

Graur, Mina. 1997. *The Anarchist "Rabbi": The Life and Teachings of Rudolf Rocker.* New York: St. Martin's Press.

Griffin, Richard. 2002. "Reality Check." *Total Liberty* 3: 5–6.

Guérin, Daniel, ed. 1998. *No Gods, No Masters.* Oakland: AK Press.

Hobsbawm, E.J. 1990. *Nations and Nationalism since 1780.* Cambridge: Cambridge University Press.

Holloway, John. 2005. *Change the World without Taking Power.* London: Pluto Press.

Lehning, Arthur. 1973. *Michael Bakunin: Selected Writings.* London: Cape.

Llobera, Josep R. 1994. *The God of Modernity.* Oxford: Berg.

Purkis, Jon, and James Bowen. 1997. *Twenty-First Century Anarchism: Unorthodox Ideas for the New Millennium.* London: Cassell.

Rocker, Fermin. 1998. *The East End Years: A Stepney Childhood.* London: Freedom Press.

Rocker, Rudolf. 1924. *Johann Most: Das Leben Eines Rebellen.* Berlin: Der Syndicalist.

_____. 1937. *Nationalism and Culture* (1978 edition). St. Paul, MN: Coughlin.

_____. 1938. *Anarcho-Syndicalism* (1989 edition). London: Pluto Press.

_____. 1949. *Pioneers of American Freedom.* Los Angeles: Rocker Publications.

_____. 2005. *The London Years.* Oakland: AK Press.

Souchy, Agustín. 1992. *Beware Anarchist! A Life for Freedom.* Chicago: Charles H. Kerr.

Woodcock, George, ed. 1977. *The Anarchist Reader.* London: Fontana.

11

The Political Legacy of Murray Bookchin (2009)

Ever since I read *Post-Scarcity Anarchism*, some thirty years ago, I have been a fan of Murray Bookchin—in the same way as I have been a fan of Peter Kropotkin, Richard Jefferies, Elisée Reclus, and Ernest Thompson Seton. All were pioneer ecologists. In 1981 in a review of a book on eco-philosophy, I described Bookchin as a "lone voice crying in the wilderness," and even ten years later still felt the need to publish an essay on "The Social Ecology of Murray Bookchin" (1996, 131–38), emphasising Bookchin's seminal importance as a social ecologist and as a radical political thinker. However, by the end of the decade, Bookchin's trenchant (and valid) criticisms of deep ecology, anarcho-primitivism, and the bourgeois individualism of the likes of Hakim Bey (aka Peter Lamborn Wilson), had thrust Bookchin into the media limelight, and he became something of a controversial figure. He certainly ruffled many feathers, especially amongst those happily ensconced in the academy. He thus came to be assailed from all sides— by deep ecologists, political liberals, technophobes, spiritual ecologists, anarcho-primitivists, poetic terrorists, neo-Marxists, and Stirnerite individualists, as well as the acolytes of Nietzsche and Heidegger.

In the process, of course, Bookchin's seminal importance as a social ecologist and as a radical anarchist thinker tended to be forgotten, if not completely denigrated. But what to me was important about Bookchin was that he reaffirmed and creatively developed the revolutionary anarchist tradition that stemmed essentially from Michael Bakunin, Peter Kropotkin, and Elisée Reclus. This tradition emphasised the need to integrate an ecological worldview or philosophy—what Bookchin was later to describe as dialectical naturalism—with the political philosophy offered by anarchism, that is, by libertarian socialism. This political tradition and social movement, as many have emphasised, combined the best of both liberalism, with its emphasis on liberty and individual freedom, and socialism with its emphasis on equality, voluntary associations, mutual aid, and direct action. This unity, that indeed defines libertarian socialism (or anarchism) was most succinctly expressed in the well-known maxim of Michael Bakunin: "That liberty without socialism is privilege and injustice, and that socialism without liberty is slavery and brutality" (Lehning 1973, 110).

Some forty or so years ago Murray Bookchin sensed that the social and the natural must be grasped in a new unity. That the time had come to integrate an ecological natural philosophy (social ecology) with the social philosophy based on freedom and mutual aid (anarchism or libertarian socialism). This unity was essential, he argued, if we were to avoid an ecological catastrophe. What we must, therefore do, Bookchin stressed, was to "decentralize, restore bioregional forms of production and food cultivation, diversify our technologies, scale them to human dimensions, and establish face-to-face forms of democracy," as well as to foster a "new sensibility toward the biosphere" (1980, 27).

Although in later years Bookchin became embroiled in rather acrimonious debates with deep ecologists, anarcho-primitivists and bourgeois individualists—in which Bookchin fervently defended his own brand of social ecology and libertarian socialism—Bookchin never, in fact, deviated from the views he expressed in his earlier writings. Bookchin's core ideas on social ecology, libertarian socialism, and libertarian municipalism—which he defended and elaborated upon throughout his life—are thus to be found in three key early texts, namely, *Post-Scarcity Anarchism* (1971), *Toward an Ecological Society* (1980), and his magnum opus *The Ecology of Freedom* (1982). As Tom Cahill remarked in his generous tribute to Bookchin, these books contain the "essence" of Bookchin's thoughts (2006, 164).

It has to be recognised that although Bookchin always expressed his views with some stridency, even rancour—to a degree that many found disturbing—he was in fact no more doctrinaire, sectarian, and ideological than the anarcho-primitivists and the individualist anarchists with whom he disputed, and he expressed a much broader social vision. What could be more narrow and sectarian than the kind of anarcho-primitivism expressed by Bob Black and John Zerzan? An Oxford University academic like Uri Gordon, deeply offended by Bookchin's "vituperative attacks" on the "new anarchists," thus comes to completely ignore the substance of Bookchin's critique (2008, 26), for anyone who has read, for example, Hakim Bey's esoteric writings can easily understand why Bookchin described them as "narcissistic," "elitist," "petit-bourgeois," and as a "credo for social indifference" (1995, 20–26). Benjamin Franks is of the same opinion. For Franks suggests that Bey's kind of bourgeois politics completely fails to confront the oppressive power of both the state and capital, happily co-existing with them, and is essentially a form of liberalism, akin, he even suggests, to anarcho-capitalism (2006, 266–67). And contrary to what many academics think, the anarcho-capitalism of the likes of Ayn Rand—Aynarchism, as Ruth Kinna (2005, 25) describes it—is by no stretch of the imagination anarchism as Bookchin described it (see my critique of Rand's politics 1996, 183–92). Bey is just an old-fashioned liberal with a penchant for Nietzschean aesthetics and Islamic mysticism, and his liberal politics were rightly condemned by Bookchin.

What Bookchin describes and critiques as "lifestyle" anarchism is in fact what many academics have now come to describe as the "new anarchism" (e.g., Kinna 2005, Curran 2006). According to Ruth Kinna (2005) this "new anarchism" consists of a rather esoteric pastiche of five ideological categories—for Bookchin can in no sense be described as a "new" or "lifestyle" anarchist! These categories are: the anarcho-primitivism associated with Bob Black and John Zerzan; the "poetic terrorism" of Hakim Bey and John Moore who follow the aristocratic aesthetic nihilism of Friedrich Nietzsche; Stirnerite individualism; the anarcho-capitalism of Murray Rothbard and Ayn Rand; and, finally, the so-called postmodern anarchism that is derived from the writings of Deleuze, Foucault, Derrida, and Lyotard. None of this "new anarchism" is in fact either new or original. What they have in common is the kind of radical individualism and neo-romanticism that Bookchin identified and critiqued as "lifestyle" anarchism.

In their response to Bookchin's critique, Bob Black, David Watson, and surprisingly, John Clark (aka Max Cafard, who at one time was a fervent devotee of Bookchin) all harshly denounce Bookchin's social ecology, and were more than a match for Bookchin in their invective. Bookchin thus came to be depicted by these three as an aspiring "anarchist Lenin," an "anarcho-leftist fundamentalist," a dogmatic "technocrat," and advocate of "spontaneous violence" due to Bookchin's "revolutionary fantasies," the arrogant promoter of some "Faustian project," as well as being described as an intellectual buffoon. Bookchin's defence of reason and truth—as against religious dogma, mysticism, and postmodern relativism—implied, it was argued, that he had affinities to the American neoconservatives, advocates of free market capitalism (Watson 1996, Black 1997, Clark 1998).

Although Robert Graham (2000) has little sympathy with the acrimonious and denunciatory polemics that have marred the anarchist debates around social ecology—and rightly so—he nevertheless defends Bookchin's integrity, and suggests that the three critics have seriously misjudged, or wilfully misinterpreted, Bookchin's social ecology.

In the bookshops now is a useful little book entitled *Social Ecology and Communalism* (2007). In many ways it constitutes Bookchin's last testament, and provides a good introduction and summary of his political legacy. It consists of four essays written in the last decades of his life, and has a short but useful introduction by the editor Eirik Eigland.

The first essay, "What Is Social Ecology?," originally published in 1993, essentially outlines Bookchin's thoughts on the emergence of hierarchy and capitalism, and his conception of an ecological society. For Bookchin, human life is essentially a paradox. For on the one hand, humans are intrinsically a part of nature, the product of an evolutionary process. That humans are conceived as "aliens" or as "parasites" on earth, as suggested by some deep ecologists and eco-phenomenologists, Bookchin found quite deplorable. It implies, he argued, a "denaturing of humanity," and denies the fact that humans are "rooted" in biology and evolutionary history.

On the other hand, in the course of their development as a unique species-being, humans have developed language, a potential for subjectivity and flexibility, and a "second nature," such that their cultures are rich in experience and knowledge. This gives humans technical foresight, and the capacity to creatively refashion their environment (24–27).

To understand the natural world as an evolutionary process, and the place of humans within the cosmos, Bookchin therefore argues that we need to develop an organic way of thinking, one that is dialectical and processual, rather than instrumental and analytic. Such a way of thinking avoids the extremes of both anthropocentrism, exemplified by Cartesian metaphysics, which radically separates humans from nature, and biocentrism, which is a naïve form of biological reductionism expressed by both deep ecologists and sociobiologists (27–28).

Early human societies, Bookchin argued, were essentially egalitarian, practising mutual aid, and following the principles of usufruct and the irreducible minimum—the notion that everyone in a community was entitled to a basic livelihood (37). Bookchin goes on to suggest that the first forms of hierarchy were based on age and gender and that it is therefore important to make a distinction between hierarchy as a form of domination and class exploitation (36).

Although the idea of dominating nature is almost as old as that of hierarchy itself, Bookchin emphasises that the current ecological crisis has its roots not in overpopulation, technology, or human nature, but in the capitalist system, which is inherently anti-ecological. It is well to recall that over forty years ago Bookchin was reporting in detail the environmental and health costs of pesticides, food additives, chemicalised agriculture, pollution, urbanisation, and nuclear power. He was even, with some prescience—long before Al Gore and George Monbiot—highlighting the problems of global warming—that the growing blanket of carbon dioxide would lead to destructive storm patterns, and eventually the melting of the ice caps and rising sea levels (1971, 60). But the cause of this ecological crisis, for Bookchin, was not because humans were inherently the most destructive parasite on earth; rather it was due to a capitalist system that was in its very essence geared to exploitation, competition, and ruthless economic expansion. This is spelled out in the second essay, "Radical Politics in an Era of Advanced Capitalism," where Bookchin describes capitalism as an "ecological cancer," a form of "barbarism" that is making the earth virtually unsuitable for complex forms of life (56). Equally important, for Bookchin, capitalism is not simply an economic system that is polluting and ravaging the natural world; it is also leading to the expansion of commodity relationships into all areas of social and cultural life. One thing that can be said about Bookchin is that he is a fervent anti-capitalist, in ways that media radicals like Naomi Klein and George Monbiot are most certainly not. For both Klein and Monbiot are simply reformist liberals, with a vision of some benign forms of capitalism.

This leads Bookchin to advocate the creation of an "ecological society," involving the following: the social transformation of society along ecological lines; the elimination of class exploitation and all forms of hierarchy and domination; a spiritual renewal that develops humanity's potential for rationality, foresight, and creativity; and the fostering of an ecological sensibility and what Bookchin describes as an "ethics of complementarity" (46–47). But crucial to Bookchin's vision of an ecological society is the need to develop a radical form of politics based on the municipality.

Unlike Nietzschean "free spirits" and Stirnerite individualists, who in elitist fashion rely on other mortals to provide them with the basic necessities of life, Bookchin recognised that throughout human history some form of social organisation has always been evident. For humans are always intrinsically social beings. Some kind of organisation has therefore always been essential, not only in terms of human survival, but specifically in terms of the care and upbringing of children (kinship), in the production of food, shelter, clothing, and the basic necessities of human life (the social economy) and finally, in the management of human affairs, relating to community decisions and the resolution of conflicts (politics). Bookchin, therefore, has always been keen to distinguish between ordinary social life—focused around family-life and kinship, affinity groups, and productive activities—and the political life of a community, focused around local assemblies.

Bookchin has been equally insistent on distinguishing between politics—which he defined as a theory relating to the public realm, and to those social institutions by means of which people democratically managed their own community affairs, and what he called "state craft." The latter was focused on the state, defined as a form of government that served as an instrument for class exploitation and for class oppression and control (95). Thus Bookchin saw "government"—institutions which deal with the problems of orderly social life—as consisting of two forms: as the state or as local democratic assemblies centred on what he described as municipal politics.

But even in his earliest writings, reflected in the seminal essay "The Forms of Freedom," Bookchin was concerned with exploring what "social forms" were most consistent with the "fullest realization of personal and social freedom" (1974, 143). It is of interest that in this early essay Bookchin is critical of the limitations of workers' councils and does not in fact use the term "government," only that of "self-management." He also indicated the dangers of an assembly becoming an "incipient state" (168).

In his last essays, however, Bookchin argues that we need a new politics based on what he describes as the "communalist project." As in the early writings, he describes the various forms of popular assemblies that have emerged throughout European history, particularly during times of social revolution. Bookchin is particularly enthusiastic about the classical Athenian polis, where citizens (aristocratic males) managed the affairs of the community through a form of direct democracy, instituted in a popular assembly. Even though, as Bookchin always recognised and stressed, such a form of democracy was marred by patriarchy, slavery, and class rule (49), the Athenian polis was in fact a city-state. But such forms of popular democracy had been found from earliest times, and Bookchin cites, for example, the popular assemblies of medieval towns, the neighbourhood sections formed during the French Revolution, the Paris Commune of 1871, the workers' soviets during the Russian Revolution, and the New England town meetings (49).

Bookchin thus comes to put a focal emphasis on the need to establish popular democratic assemblies, based on neighbourhoods, towns and villages. Such local assemblies through face to face democracy, would make policy decisions relating to the management of community affairs (101). He argues consistently that such decisions should be made by majority vote, though Bookchin does not advocate majority rule (109), and emphasises that a free society would only be one that fosters the fullest degree of dissent and liberty. He is, however, given his early experiences with the anti-nuclear Clamshell Alliance, highly critical of consensus politics, except for small groups (110).

But Bookchin goes on to argue that such local or municipal assemblies must be formally structured, with constitutions and explicit regulations (111), and that the assembly, as the sole policy-making body, has priority over the workers' committees and the cooperatives concerned with food production and other social activities. These would have a purely administrative function. As Bookchin puts it: "every productive enterprise falls under the purview of the local assembly, which decides how it will function to meet the interests of the community as a whole" (2007, 103). Town and neighbourhood assemblies would be linked through con-federal councils, consisting of mandated delegates sent by the assemblies (50). It seems important for Bookchin that power be both decentralised, and instituted in local communities, organised through face-to-face demo-cratic assemblies. Even more controversial, Bookchin advocates that com-munalists (i.e., libertarian socialists) should not hesitate to run candidates in local government elections, and thereby attempt to convert them to popular assemblies (115).

What has troubled many anarchists is that while the "lifestyle" or "new" anarchists (whether anarcho-primitivists, poetic terrorists, poststructur-alist anarchists, or Stirnerite egoists) have, as ultra-individualists, deni-grated, or even repudiated the socialist component of anarchism—derided as "leftism" (that is, they have repudiated political protest and class strug-gle)—Bookchin in his later years, partly in reaction to the "lifestyle" anar-chists, has moved to the other extreme and has increasingly downplayed not only cultural protest but the libertarian aspect of anarchism. Thus his emphasis on local assemblies and confederations as structured insti-tutions that take priority not only over voluntary associations and self-management of the economy, but also, it seems, over the individual, seems to many to introduce an element of hierarchy quite foreign to anarchism, that is, libertarian socialism or anarchist communism. In fact, the whole idea of "government" seems contrary to anarchist principles.

Bookchin has always acknowledged the importance of protests and struggles to achieve a better world—whether centred around nuclear power, ecological issues, health care and education, or community issues, as well as the importance of the anti-globalisation movement in challeng-ing capitalism, both on cultural and economic grounds (85). Nevertheless, he has tended to focus "direct action" rather narrowly on local municipal elections. This also seems contrary to libertarian socialist principles, for local authorities are essential appendages of the nation-state. This strategy is thus basically reformist.

Bookchin's critique of "lifestyle" or "new" anarchism is, I think, largely justified and valid. In fact, the essay "The Role of Social Ecology in a Period of Reaction" is largely devoted to a reaffirmation of what was expressed in his controversial polemic *Social Anarchism or Lifestyle Anarchism* (1995). For besides emphasising that social ecology is deeply rooted in the ideals of the radical Enlightenment and the revolutionary socialist tradition (71), Bookchin argues that the "new" or "lifestyle" anarchism, as expressed by the likes of Hakim Bey, Bob Black, and Jason McQuinn, is largely a retrogressive "goulash" in its embrace of spiritualism, anti-rationalism, primitivism, and bourgeois individualism. Lifestyle anarchism, he writes, with some derision, is little more than an ideology that panders to petit bourgeois tastes in eccentricity (72).

Thus his hostility towards "lifestyle" anarchism and radical individualism, combined with his advocacy of a highly structured form of municipal "government" (no less) has led Bookchin to almost forget the libertarian component of anarchism and the cultural importance of the concepts of individual freedom and autonomy, both personal and social, as well as of cultural revolt. Indeed, in his early writings Bookchin put a crucial emphasis on the self, on self-activity and self-management, arguing that a truly free society does not deny selfhood and individual freedom, but rather supports and actualises it (1980, 48). He even advocates lifestyle politics as being an indispensable aspect of the revolutionary project (1974, 16). But as Robert Graham (2004) has argued, Bookchin's later writings on "communalism," with its focus almost exclusively on the structured municipal assembly, tends to downplay or marginalise direct action, the self-management of the economy, and the crucial importance of individual freedom. Anarchism has a dual heritage, and must not only be socialist (denied by most of the "new" or lifestyle anarchists) but also libertarian—which seems to be rather downplayed by Bookchin in his last years.

It has to be recognised, of course, that although Bookchin is highly critical of Marxism and the idea of a "proletarian revolution," as well as of anarcho-syndicalism given his hostility to the "factory system," Bookchin never repudiated the concept of the "class." He always acknowledged—as a fervent anti-capitalist—the crucial importance of the working class in achieving any form of social revolution, and categorically affirmed the importance of class struggle (1999, 264).

It is also important to note that although Bookchin was a harsh critic of the kind of anarcho-primitivism that essentially stemmed from the writings of Fredy Perlman, he was not an obsessive "technocrat" as portrayed by Watson (1996)—in fact Bookchin described himself as a bit of a Luddite. Nor was he besotted with civilisation. He certainly emphasised the importance of the city, especially in introducing the idea of a common humanitas (61); but like both Peter Kropotkin and Lewis Mumford—both important influences on Bookchin—and unlike the anarcho-primitivists, Bookchin had a much more nuanced approach to both technology and civilisation. As he put it, in defending his pro-technology stand: "which is not to deny that many technologies are inherently domineering and ecologically dangerous, or to assert that civilisation has been an unmitigated blessing. Nuclear reactors, huge dams, highly centralized industrial complexes, the factory system, and the arms industry—like bureaucracy, urban blight, and contemporary media—have been pernicious almost from their conception" (1995, 34).

Following Kropotkin, Bookchin therefore came to emphasise that there had been two sides to human history—a legacy of domination reflected in the emergence of hierarchy, state power, and capitalism; and a legacy of freedom, reflected in the history of ever-expanding struggles for emancipation (1999, 278).

It is thus disheartening to read, in the last essay, "The Communalist Project," that Bookchin comes to deny that he is an anarchist; that he had embraced, as an alternative, the politics of "communalism." Rather ironically, communalism is defined as a form of libertarian socialism, and is seen as the political dimension of social ecology, libertarian municipalism being its praxis (108).

Significantly, making clear demarcations between Marxism, anarcho-syndicalism, and anarchism, Bookchin comes to narrowly define anarchism in terms only of its individualistic tendency. Thus in both the essay, and in his preface to the third edition of *Post-Scarcity Anarchism* (2004), Bookchin comes to define anarchism as a "tangle of highly confused individualistic concepts." Anarchism is thus misleadingly interpreted in terms of "lifestyle" anarchism, characterised by ultra-individualism, nihilism, mutualism, aestheticism, and as being radically opposed to any form of organisation. Both conceptually and historically this is an inaccurate depiction of anarchism, which has always embraced a dual heritage of liberty and socialism. But it leads Bookchin—like the Marxists, anarcho-primitivists, and Stirnerite egoists—to postulate a false and quite untenable dichotomy between anarchism and socialism. For historically the main strand of anarchism has been anarchist communism (or libertarian socialism) combining liberalism—as existential not possessive, individualism—with socialism. The socialism that Bookchin now espouses as communalism, which he affirms as both libertarian and revolutionary (96), is in fact good old-fashioned anarchism. First formulated by Bakunin towards the end of the nineteenth century, anarchism in this sense has various synonyms: anarchist communism, revolutionary anarchism, libertarian communism, class struggle anarchism, or as Bookchin and many contemporary anarchists conceive it: social anarchism or libertarian socialism.

Authentic anarchism is not then the lifestyle (or "new") anarchism—as Bookchin contended in his last years—but the class struggle anarchism embraced by Reclus, Kropotkin Goldman, Berkman, Flores Magón, Galleani, Malatesta, Landauer, and by scores of contemporary anarchists and radical activists who muster (at least in Britain) under such banners as Class War, the Solidarity Federation (the Direct Movement), Black Flag, Industrial Workers of the World, and the Anarchist (Communist) Federation (see

Franks 2006). Bookchin, in spite of his rhetoric, and in spite of misleadingly equating anarchism with ultra-individualism, always essentially belonged to this libertarian socialist tradition—anarchism. Bookchin's true legacy, it seems to me, was in reaffirming and creatively developing this tradition, not in advocating libertarian municipalism, with its rather reformist implications.

References

Black, Bob. 1997. *Anarchy After Leftism*. Columbia, MO: CAL.

Bookchin, Murray. 1974. *Post-Scarcity Anarchism*. London: Wildwood House.

_____. 1980. *Toward an Ecological Society*. Montreal: Black Rose.

_____. 1995. *Social Anarchism or Lifestyle Anarchism*. San Francisco: AK Press.

_____. 1999. *Anarchism, Marxism and the Future of the Left*. San Francisco: AK Press.

_____. 2007. *Social Ecology and Communalism*. Oakland: AK Press.

Cahill, Tom. 2006. "Murray Bookchin (1921–2006)" *Anarchist Studies* 14, no. 2: 163–66.

Clark, John. 1998. "Municipal Dreams: A Social Ecological Critique of Bookchin's Politics." In *Social Ecology after Bookchin*, edited by Andrew Light. New York: Guildford Press.

Curran, Giorel. 2006. *21st Century Dissent: Anarchism, Anti-Globalization and Environmentalism*. Basingstoke: Palgrave.

Franks, Benjamin. 2006. *Rebel Alliances*. Oakland: AK Press.

Gordon, Uri. 2008. *Anarchy Alive!* London: Pluto Press.

Graham, Robert. 2000. "Broken Promises: The Politics of Social Ecology Revisited." *Social Anarchism* 29: 26–41.

_____. 2004. "Reinventing Hierarchy: The Political Theory of Social Ecology." *Anarchist Studies* 12, no. 1: 16–35.

Kinna, Ruth. 2005. *Anarchism: A Beginner's Guide*. Oxford: Oneworld.

Lehning, Arthur, ed. 1973. *Michael Bakunin: Selected Writings*. London: Cape.

Morris, Brian. 1996. *Ecology and Anarchism: Essays and Reviews on Contemporary Thought*. Malvern Wells: Images.

Watson, David. 1996. *Beyond Bookchin: Preface for a Future Social Ecology*. Brooklyn: Autonomedia.

12

Kropotkin and the Poststructuralist Critique of Anarchism (2009)

1. Prologue

Over the past decade or so, academic scholars, along with anarcho-primitivists, Stirnerite individualists, and autonomous Marxists, have been asserting to contemporary radical activists that class struggle anarchism is now "obsolete" or "outmoded" or in need of a "major overhaul" (Black 1997; Purkis and Bowen 1997, 3; Kinna 2005, 21; Holloway 2005, 21).[1]

By "anarchism" they appear to mean the social or class struggle anarchism that derives from Bakunin and Kropotkin and has been embraced by generations of activists throughout the twentieth century, from Goldman, Rocker, and Landauer in the early part of the century, to Murray Bookchin, Colin Ward, and the Anarchist (Communist) Federation in more recent decades.[2]

We are thus informed that a "new anarchism" or a "new paradigm" has emerged that has completely replaced the "old" class struggle anarchism. According to Ruth Kinna (2005, 21–37) this "new anarchism" consists of a rather esoteric pastiche of several political tendencies, namely: the anarcho-primitivism of John Zerzan (1994); the anarcho-capitalism of Murray Rothbard and Ayn Rand; the "poetic terrorism" that derives from Nietzsche and which has been embraced with fervour by Hakim Bey (1991) and John Moore (2004) as the new "ontological anarchy"; acolytes of the possessive individualism (egoism) proclaimed by the left-Hegelian Max Stirner; and finally, the poststructuralist anarchism advocated by Todd May (1994), which is derived from the writings of the French philosophers Jacques Derrida, Gilles Deleuze, Michel Foucault, and Jean-Francois Lyotard. None of whom, it is worth noting, were anarchists.

I have elsewhere offered some critical reflections on this so-called "new anarchism," questioning whether there is anything particularly "new" about it. For it is largely a reaffirmation of nineteenth-century bourgeois individualism. I also emphasised that social or class struggle anarchism (libertarian socialism) is still a vibrant and ongoing radical movement and political tradition, and one very much involved in contemporary struggles and protests (see Sheehan 2003, Franks 2006).

Here I want to discuss one particular strand of the "new anarchism," namely poststructuralist or postmodern anarchism, otherwise known as "post-anarchism" (Newman 2001, Call 2003, Day 2005), and to specifically focus on what has been described as the "poststructuralist critique" of anarchism.[3]

2. The Poststructuralist Critique

The "poststructuralist critique" of anarchism seems a rather strange expression, as none of the poststructuralist philosophers ever mention anarchism, let alone critically engage with the writings of anarchists like Bakunin, Kropotkin, and Elisée Reclus. Jacques Derrida's scholastic musings on the politics of Karl Marx, *Specters of Marx* (1994), for example, was essentially aimed at demonstrating his own radicalism, to affirm that he was not a nihilist or apolitical as his early writings undoubtedly suggested. Indeed Derrida emphasised that he was a philosopher and not a mystic, was motivated by an interest in reason and truth, and was critical of the whole idea of the "postmodern" (Rötzer 1995, 46–47). But in *Specters*, though there are plenty of references to Maurice Blanchot, Martin Heidegger (both pro-fascists), and Max Stirner, Derrida not only never mentions Bakunin, but utterly fails to discuss the anarchist critique of Marxism.

Likewise, *A Thousand Plateaus* (1988), which is considered to be Gilles Deleuze and Félix Guattari's most political work, is written in the most impenetrable jargon, which even their devotees have difficulty in understanding. But significantly, though expressing some anarchist ideas in their "treatise on nomadology" (in a discussion of "the war machine" against the "state apparatus") apart from a brief mention of the anthropologist Pierre Clastres, there is no mention at all in the book of anarchism, let alone critically engaging in the work of any anarchist.

Thus neither Derrida, nor Deleuze, nor Lyotard, nor Foucault, all rather detached philosophical mandarins, seem to have expressed any interest in the writings of Bakunin, Reclus, Kropotkin, or any other anarchist, let alone produced a critique. Nor do they express any interest in the work of their contemporary, Murray Bookchin. Still less did they engage with the French anarchist movement, at least in any radical sense. In fact, Lyotard ended up dreaming of intergalactic travel and supporting the right-wing Giscard d'Estaing in the French presidential election, while

Foucault became an apologist for the reactionary Islamic clerics at the time of the Iranian revolution and came to renounce all aspirations for a new social order (Eagleton 2003, 37).[4]

Nevertheless, we are told that the poststructuralists (or postmodernists) have presented an important critique of social or class struggle anarchism.

So let us now discuss this so-called poststructuralist critique of anarchism—and by anarchism contemporary critics essentially mean the earlier social anarchism of Bakunin, Kropotkin, Goldman, Malatesta, Berkman, and Rocker. This critique is well expressed by several scholars, all of whom have become rather enchanted with poststructuralism or postmodernism (see for example, May 1994; Morland 2004; Call 1999; 2003; Newman 2001; 2004).

This critique asserts the following:

- that early anarchists were simply and narrowly anti-statist, and thus failed to recognise or challenge other forms of power and oppression. Untrue!
- that anarchists viewed "power" only as repressive or coercive and this did not acknowledge that "power" was also "productive." Untrue!
- that anarchists have a Cartesian notion of human subjectivity or an "essentialist" conception of human nature which they view as fixed, immutable and essentially benign. Untrue!
- that old anarchists were obsessed with workers' power and the class struggle and therefore lacked any ecological sensibility. Untrue!
- that anarchists uncritically embraced the Enlightenment project and thus acclaimed reason, progress, humanism, and science, to the neglect of culture, poetry, the arts, the emotions, and the imagination. Untrue!
- and finally, that the old anarchists were "ideological" radicals, in that they explicitly framed their aims and the ethical and political principles that guided their actions. True!

Let me take each of these six criticisms in turn, focusing on the social anarchism of Peter Kropotkin, although it must be said that they are less criticisms than serious misrepresentations of the radical ideas of an earlier generation of anarchists.

3. Anti-Statism

That Kropotkin and other anarchists were only anti-state theorists and failed to recognise or challenge other forms of power and oppression, borders on being a rather silly accusation. But it has been expressed by a host of radical academics opposed to class struggle anarchism, typical examples being John Moore (1998), L. Susan Brown (1993, 157), and Saul

Newman (2004). Although it misrepresents Kropotkin's own critique of power, it does seem to satisfy scholars eager to promote their own originality. For example, Kropotkin was suggesting a "maximalist" (ugh!) critique of power long before John Moore, without ever adopting Moore's bourgeois project of an aesthetic "ego" at "war" (no less!) with society.

For a start, Kropotkin was not only anti-capitalist, but opposed to all forms of economic exploitation—whether in the Siberian salt mines, the capitalist factories, or in relation to Russian serfdom. He also recognised that under the guise of the modern state, power relations had encroached into all aspects of social life and, as Richard Day recognised, Kropotkin, long before Habermas, Foucault, and Mario Tronti, was expressing the "colonization thesis" that the state is taking over, capturing, colonising existing social relations and putting them to work in the name of its own authority (2005, 144). As Kropotkin himself put it, "today the state takes upon itself to meddle in all areas of our lives" (1885, 25). Thus the modern state has increasingly been intruding into all aspects of social life, taking over functions—education, health, social welfare, leisure, recreation—that had earlier been organised through voluntary associations and local communities. Kropotkin, of course, critiqued and challenged all these intrusions of state power.

Long before Foucault, Kropotkin also made a distinction between the state and government. The state implied the concentration of power in the hands of a political elite, the control of a specific territory, and domination through the state apparatus—army, police, and the administrative and judicial functionaries. In contrast, government was a more general concept, reflecting the forms and mechanisms of power that involved the management and policing of the working class (1993, 163). Like Foucault, Kropotkin always emphasised that the modern state, including the ubiquitous "mechanisms" of power, played an important role in supporting capitalist hegemony.

Kropotkin also vehemently opposed state schooling and the prison system. However, unlike Foucault and Deleuze, Kropotkin was concerned not simply with prison "reform" but abolishing the prison system altogether. For Kropotkin, prisons were simply "obsolete" (see Davis 2003).

Equally important, Kropotkin was critical of all forms of social power and all ideologies—economic, political, religious—that undermined or curtailed the freedom of the individual. He was thus critical of all religious ideologies—emphasising the close relationship that had always existed between spiritualism and political hegemony—whether involving the tribal shaman or religious institutions like the Catholic Church. Like Marx, he was critical of liberal economic theory and Stirner's individualism, emphasising their ideological function, as well as being opposed to the Social Darwinism that was popular at the end of the nineteenth century.

Thus Kropotkin did not naively view the state as the "root of all evil," as Susan Brown herself assumes (1993, 157), but was opposed to all forms of coercive power and authority. He was thus critical of both private property and the wage system—which he described as "wage slavery"—and all forms of religious authority, not just the state.

Needless to say, Kropotkin was a member of the International Working Men's Association, and like Bakunin, was highly critical of Marx's party dictatorship and the hierarchical relationships developing within the socialist movement itself. Long before the autonomist Marxists Kropotkin was criticising the notion of the vanguard party, or rule by intellectual savants.

Like many of his nineteenth-century contemporaries Kropotkin lacked a gender perspective, but it is clear from his discussion of Bakunin in his *Memoirs* that both he and Bakunin were opposed to any form of gender inequality (1889, 289). Kropotkin was also particularly critical of many religious communes, which he felt had made women the "slaves" of the community, reducing them to the role of domestic servants (1997, 17).

To view Kropotkin (and other social anarchists) simply as naïve anti-statists, or as conceiving "power" as residing solely in the state, is thus seriously misleading. For Kropotkin was critical not only of all manifestations and techniques of political power, but all ideologies, relationships, and institutions that limited or inhibited the autonomy and self-development of the individual—Brown's existential individual no less! But unlike Brown, Kropotkin, as an anarchist communist, recognised and stressed that "true" individualism could only be expressed and accomplished in a society free of all hegemonic structures—a society which he described as "free communism" (for critiques of Brown's existentialist individualism see Bookchin 1995, 13–18; Morris 2004, 184–86; and for her defence of radical humanism see Brown 2004).

We may thus conclude that it is quite erroneous to dismiss Kropotkin as merely "anti-statist," for he stood firmly in the social anarchist tradition that expressed an opposition or negation of all hierarchical power relations—political, economic, social, and cultural. He was thus critical of all forms of domination, sovereignty, representation, and hierarchy (Wieck 1979, 138–39). He would have agreed with Rudolf Rocker that "common to all (social) anarchists is the desire to free society of all political and social coercive institutions that stand in the way of the development of free humanity" (1989, 20).

4. The Productivity of Power

The second criticism, namely that Kropotkin and other "old" anarchists failed to recognise that "power" is "productive" and not just repressive (May 1994, 63; Patton 2000, 8), is also completely misguided.

For the past two hundred years social scientists, political theorists, and anarchists have all recognised that no hegemonic power or political ruler, even the most bloodthirsty tyrant, rules solely by means of repression and coercion. This was recognised long ago before Foucault arrived on the intellectual scene, as was the close relation between power and knowledge. Thus "power" has always been "productive" and throughout history there has been a symbiotic relationship between political domination and religious ideologies. Biopower is nothing new! Thus anarchists like Kropotkin were not so dumb as to fail to recognise that "power" (in the sense of *pouvoir*, "power over" encapsulated in social institutions as opposed to *puissance*, potentiality or creative power) is "productive" in that it produces prisons, propaganda, disciplined subjects, laws, ideologies, festivals, and forms of knowledge, as well, of course, modes of resistance. But for Kropotkin, as for Foucault, this "productivity" is not to be interpreted as something necessarily wholesome or conducive with human well-being. Far from it: it essentially refers to techniques of social control and social regulation that is achieved through non-coercive means (Morris 2004, 209). Kropotkin was, of course, critical of such forms of power, and in contrast advocated the kind of power expressed in direct action and in the creation—the "production"—of new and alternative forms of social cooperation through voluntary associations and mutual aid. Power, as Bakunin expressed it, may be just as beneficial as harmful: "It is beneficial when it contributes to the development of knowledge, material prosperity, liberty, equality, and brotherly solidarity, harmful when it has opposite tendencies" (Lehning 1973, 150).

The notion that Kropotkin and the early anarchists saw power only as repressive or coercive is thus quite fallacious.

5. Essentialism

One of the most banal, misleading and oft-repeated criticisms of class struggle anarchists such as Kropotkin is that they hold an essentialist view of the human subject.

Everyone throughout the world, and in all cultures, express in their thoughts and actions some conception of human nature, and academic philosophers, Stirnerite egoists and Nietzschean aesthetes are no exception. However, in critiquing Kropotkin and other anarchists for holding essentialist conceptions of human nature, the implication is twofold: either they—the anarchists—posit the human person as having a fixed, immutable, benign metaphysical essence (May 1994, 63–64; Patton 2008, 8; Newman 2004), or that they articulated a Cartesian conception of human subjectivity (Call 1999, 100).

It seems to me that such postmodernists completely misrepresent the views expressed by Kropotkin and other anarchists on human subjectivity.

For a start, unlike their poststructuralist detractors, Kropotkin (like Elisée Reclus) was an evolutionary thinker, and thus recognised that humans as a species-being are the product of a long evolutionary history and development. Their sociality is therefore not some metaphysical essence—as Newman (2004, 113) supposes—but a product of evolution. Kropotkin also recognised, like Marx and Bakunin before him, that human beings were social beings, not Cartesian rational monads, nor the "abstract" individual of bourgeois ideology—the asocial, possessive, power-seeking individual of Hobbesian theory (MacPherson 1962; Morris 1994, 15–18). Marx, Bakunin, and Kropotkin all critiqued—indeed ridiculed—these "abstract" conceptions of the human person long before Lacan and the poststructuralists.

When, for example, Marx (and Engels) in *The Communist Manifesto* called on workers of all countries to unite, this was because he recognised that all workers had multiple identities—in terms of gender, race, nationality, and occupation, and that the "man" of bourgeois theory had no reality, that it was an abstraction, that it existed only in "the misty realms of philosophical fantasy" (1968, 57). That is why in his early writings Marx defined the human person as an "ensemble of social relations."

That humans are social beings and not disembodied rational egos or "abstract" (bourgeois) asocial individuals was thus clearly recognised by both Bakunin and Kropotkin, as well as by generations of social scientists and anarchists, at least since Marx.

What happened was that in the 1970s—in opposition to Sartre's existentialism and Husserl's phenomenology—French academic philosophers suddenly discovered for themselves what had been common knowledge among social scientists for more than a century. Namely, that human beings are social beings, not Cartesian monads, and that self-identity— personhood—in all human cultures is complex, embodied, shifting, relational, and involves multiple identities.

Thus, it has to be said, sociobiologists, evolutionary psychologists, behaviourists, Stirnerite egoists, cultural aesthetes like Hakim Bey and John Moore, and even poststructuralist philosophers like Foucault and Deleuze, are much more prone to essentialist thinking than Kropotkin ever was.

Indeed, it is of interest to note that Kropotkin suggested that the conception of an "abstract" asocial individual—whether the Cartesian rational monad, or the Hobbesian (Stirnerite) possessive individual—was a recent conception in human history. It was, he felt, intrinsically linked to emergence of capitalism, to the treatment of human labour as a commodity. It was thus totally lacking in tribal society. For tribal peoples had a sociocentric conception of the human subject; that is, a non-essentialist concept of human nature (Morris 2004, 177–90).

Equally interesting is that both Foucault and Moore seem to express a rather Hobbesian (that is, an essentialist) conception of the human person, in viewing interpersonal relationships as implying "war"—an inherent war of "all against all," that intrinsically, we "all fight each other" as Foucault put it (1980, 208, Moore 1998, 40). Neither appears to have read Kropotkin's *Mutual Aid* (1902).

Also important to note is that Kropotkin did not view the person as possessing some benign, metaphysical essence—he was as critical of Rousseau as he was of Hobbes—or as ever completely divorced from social and political relations. Thus in foisting upon social anarchism a "Manichean logic," that is, an absolute dichotomy between good and evil, Newman (2004, 109) presents a biased and quite misleading portrait of Kropotkin's conception of human subjectivity, one that—allegedly—implies an "essentialist identity." This essentialist identity is largely a figment of Newman's own imagination. For it not only depends on the conflation of several distinct conceptions of the human person (Morris 1994, 10–13), but has very little connection with Kropotkin's own identity as a social being or his conception of human nature. In fact John Moore's "new" (Stirnerite) anarchism reflects, far more than social anarchism, a "Manichean logic." For he appears to see nothing between "power"—as a totalising metaphysical abstraction that is wholly negative—and the human subject interpreted, in following Stirner and Nietzsche, as an isolated, asocial ego, anxious to assert its power in poetic insurrection.

(For a discussion of Bakunin's and Kropotkin's conception of the individual, which includes a rebuttal of their academic detractors see Morris 1993, 92–94; 2004, 180–90).

6. Ecology

Early social anarchists, obsessed with worker's control and the class struggle, were oblivious, we are told, to the environmental issues. By contrast, a key characteristic of the "new anarchism" is an ecological sensibility (Kinna 2005, Curran 2006).

Two points need to be made. First, it would be difficult to find anybody these days, whatever their politics, who does not claim to be "green" and intent on "saving" the planet earth. Second, many of the so-called poststructuralist anarchists—for example, Derrida, Baudrillard, Lyotard, and Foucault—can hardly be described as ecological thinkers, for their antirealism virtually oblates the natural world.

But, of course, this criticism ignores the fact that the two key figures in the development of class struggle anarchism—Elisée Reclus and Kropotkin—were both pioneers in the development of an ecological worldview. Both scholars were important anarchist geographers, and both developed a metaphysics of nature that completely undermined the dualism, the anthropocentrism and the determinism of the Cartesian mechanistic worldview. This was long before the environmental philosophy (Zimmerman, Callicott), deep ecology (Naess), quantum physics and systems theory (Bateson, Capra), and eco-feminism (Spretnak). For what Kropotkin (and Reclus) recognised and affirmed were the following: that humans were not the special products of god's creation but evolved according to the principles that operated throughout nature; that there is an intrinsic physical and organic (not spiritual) link between humans and nature, such that humans were an integral part of nature; and that openness, chance, creativity, and the agency and individuality of all living beings were integral aspects of the evolutionary process. Finally, Kropotkin and Reclus suggested a way of understanding that was naturalistic and historical (not static and spiritual). Thus human understanding and knowledge involved both critical reason and empirical observation and experiment.

Thus, in many ways, as Graham Purchase suggested, Kropotkin expressed in embryonic form seminal ideas that find a resonance in contemporary chaos theory and evolutionary biology. For Kropotkin, like Reclus, emphasised the importance of self-organisation, complexity, and the idea that "order" is not necessarily something externally imposed, but is implicate and emerges spontaneously (Baldwin 1927, 118–19; Purchase 1996, 138). Equally important of course, is that Kropotkin's embrace of Darwin's evolutionary theory meant that he abandoned what Ernst Mayr (2002, 74) described as typological or essentialist thinking. All this is lost on Kropotkin's recent detractors.

The dismissal of Kropotkin as a crude positivist (Crowder 1992) or as a Cartesian rationalist (Call 1999)—he could hardly be both!—thus indicates a woeful misunderstanding of Kropotkin's (and Reclus's) metaphysics of nature (see Morris 2001, 2004, 113–27; Clark and Martin 2013, 16–34).

Unlike many "new anarchists," Kropotkin (and Reclus) did not see any dichotomy, let alone an opposition, between "green" anarchism, a concern for the environment and the "rights" of nature, and class struggle anarchism and a concern for class issues and social justice. In fact, in developing the social anarchist tradition Murray Bookchin (1980, 1982) was later to strongly affirm that there was an intrinsic link between the domination of humans and the domination of nature.

Kropotkin has often been dismissed by Marxists as a utopian dreamer. He was, of course, nothing of the kind; and in *Fields, Factories and Workshops* (1899)—which Colin Ward described as one of the prophetic books of the nineteenth century—Kropotkin outlined a form of social economy which emphasised the productivity of small-scale decentralised industry, the importance of horticulture, and the need to integrate agriculture and manufacture in a decentralised economy. In doing so, Kropotkin offered important critiques of the factory system, petty commodity production and large-scale capitalist agriculture. He was thus an inspiration for both the social ecologist Lewis Mumford (1970)—in his critique of the mega-machine—and Colin Ward (1973) whose ecological anarchism emphasised "anarchy in action."

It is also worth noting that Kropotkin, in *Mutual Aid* (1902), critiqued the ultra-Darwinism espoused by Thomas Huxley and Herbert Spencer. This theory emphasised that nature always involved competitive struggle and the "survival of the fittest"—that nature was "red in tooth and claw." In contrast, Kropotkin argued that throughout the natural world, and throughout human history, mutual aid and cooperation was an important factor in evolution. Such ultra-Darwinism is still an important trend in contemporary biology, especially among such biologists as Edward Wilson and Richard Dawkins (for critiques see Morris 1991, 132–42; Rose and Rose 2000).

Thus there is no evidence at all that early anarchists lacked an ecological perspective—at least in relation to such social anarchists or libertarian socialists as Kropotkin, Elisée Reclus, Edward Carpenter, and Gustav Landauer (see for example Clark and Martin 2013, Barua 1991, Lunn 1973).

7. The Enlightenment

It is a common pastime among many postmodern academics (as well as among the "new anarchists") to express a blanket dismissal of the Enlightenment tradition. Indeed to hold the Enlightenment or reason as responsible for the political horrors of the twentieth century, rather than engaging with the real factors, namely, inter-state conflicts, fascist ideology, and the realities of an expanding capitalism. It is a bit like blaming Jesus for the Inquisition and equally misconceived.

In many ways Kropotkin, like Marx, stood firmly in the tradition of the Enlightenment, which was not, of course, a purely French phenomenon. He has therefore been dismissed by "postmodern" anarchists as an Enlightenment "rationalist," or as a "humanist," or even as a "modernist." Certainly Kropotkin embraced many of the radical aspects of the Enlightenment tradition; the affirmation of such universal values as individual liberty, equality, and fraternity; the promotion of a cosmopolitan outlook; a stress on the importance of free enquiry, secularism, and religious tolerance; an advocacy of critical reason and scientific materialism and thus the repudiation of knowledge based on mystical intuition, divine revelation, and religious dogma; and finally, a respect for craft industry and a belief in human progress through the application of scientific knowledge and technology. In this sense Kropotkin embraced "modernity" as opposed to political and religious absolutism and the authority of tradition and theology.

But it is also important to recognise that Kropotkin renounced many of the aspects of the Enlightenment tradition (or "modernity"). He repudiated, for example, not only the absolutist state but the democratic state and the whole idea of representative government. Equally important, Kropotkin rejected capitalism and the market economy, and was particularly critical of "private property," a key concept for Enlightenment liberals like Locke. Kropotkin also repudiated the metaphysics of the Enlightenment with its mechanistic conception of nature, its radical dualism and its essentialist conception of the human subject, whether as a rational monad (Descartes) or an abstract asocial individual (Hobbes, Stirner). For as noted above, Kropotkin recognised that evolutionary theory and advances in the natural and social sciences at the end of the nineteenth century had completely undermined the mechanistic philosophy of the Enlightenment. He thus came to acknowledge a "new philosophy"—which may be described as evolutionary holism (Baldwin 1927, 116–19; Morris 2004, 113–27).

Kropotkin was also critical of the moral philosophies associated with the Enlightenment, both Kant's rationalist ethics (deontology) and the utilitarianism of both John Stuart Mill and Jeremy Bentham. Kropotkin thus advocated an ethical naturalism that entailed a "prefigurative ethics," a linking of means and ends, as Benjamin Franks (2006, 94) describes it, without ever exploring Kropotkin's own theory of ethics. This, of course, completely undermined the fact/value dualism upheld by such Enlightenment thinkers as David Hume, and the later positivists.

To describe Kropotkin (and other class struggle anarchists) as "modernist" is therefore quite misleading. In many respects he was "anti-modernist."

Kropotkin embraced the concept of "reason" to undermine absolutism, and to reaffirm the crucial importance of human agency and social freedom. But this did not imply that either Kropotkin or such Enlightenment thinkers as Hume, Diderot, and Adam Smith did not put an equal stress on the importance of other human faculties such as intuition, emotions (passions) and the imagination. Following Smith, Kropotkin argued that our moral conceptions are the product of both our feelings and our reason, and have developed naturally in the life of human societies. Particularly important for Kropotkin were the feelings of sympathy and solidarity, neither of which were unique to humans, as they existed among all social animals (Kropotkin 1924, 199–208; Morris 2004, 161–63).

Because Kropotkin put an emphasis on critical reason and empirical science, this did not imply that he ignored or devalued other aspects of human life and culture. Like Emma Goldman, Kropotkin often gave lectures on literature, and was particularly well-versed in Russian literature and drama. In fact, in England Kropotkin was greatly respected in literary circles, and Oscar Wilde, Bernard Shaw, and William Rossetti were among his friends and admirers (see introduction by George Woodcock to Kropotkin 1905).

Kropotkin and other class struggle anarchists are often dismissed as an Enlightenment "humanist" (or "rationalist")—and the terms are used in the most negative and derogatory fashion. If one means by humanism (or rationalism) a belief in the capacity of humans themselves, through their reason and imagination, to deal with the problems of human existence, and thus the rejection of any spiritualist or religious metaphysics, together with the affirmation of such universal human values as solidarity and freedom, tolerance, empathy, and equality, such values being relative to humans and not derived from any divine source, then Kropotkin and other class struggle anarchists were committed humanists and rationalists.

If, however, one means by "humanism" the idea that the human species is some godlike creature, that it implies a Cartesian conception of human subjectivity, and that it also entails a Baconian (or Faustian) ethic sanctioning the human domination of nature by means of technological mastery, then, most certainly, Kropotkin was not a humanist. Neither was he a philosophical rationalist, for like Darwin he stressed the importance of empirical knowledge.

It is, then, rather disheartening to observe the "new" or postmodern anarchists joining hands with religious mystics and apologists for religion in misleadingly equating humanism (and science) with anthropocentrism and an ethic of domination.

Although Kropotkin lived at a time when the "myth of progress" was a dominant motif, the anarchist, like his friend Elisée Reclus, had a more nuanced, comprehensive and holistic conception of human history. He saw history as neither a chronicle of unending progress, nor, as with the anarcho-primitivists, as one of cultural degradation and decline after some alleged "fall" (agriculture) from grace. Thus although Kropotkin affirmed the importance of science and technology, in overcoming ignorance and material scarcity, he also acknowledged the many retrogressive aspects of human history—the rise of hierarchical relations and state power in particular. Thus Kropotkin suggested that throughout human history there had always been "two tendencies." A legacy of control and domination expressed by shamans, prophets, priests and governments, and a legacy of freedom and struggle expressed not only in the creative power of people themselves in establishing social institutions and voluntary associations, but in the struggle of people throughout history for emancipation and autonomy (Baldwin 1927, 146–47; Bookchin 1999, 278).

As right-wing neoliberals are now taking up a defence of the Enlightenment tradition, allegedly to counter Islamic fundamentalism in the "clash of civilizations," it behoves anarchists to follow Foucault in rejecting the whole idea of being for or against the Enlightenment. Thus refusing to succumb to a kind of "blackmail"—as Foucault called it (Rabinow 1984, 42–43).

The legacy of the Enlightenment is an ambiguous legacy, as Kropotkin acknowledged long ago (for a radical defence of the Enlightenment see Bronner 2004). To discuss Kropotkin as an Enlightenment "rationalist," as if this was some kind of intellectual aberration, is therefore unhelpful and obfuscating.

8. Ideology

A final criticism that is foisted on the earlier generation of class struggle anarchists like Kropotkin, as well as upon contemporary anarchists like Murray Bookchin and the Black Bloc, is that they are ideological. Rehashing the old wishy-washy liberal idea of the "end of ideology," and contending that the "new anarchists" (as well as Stirner and Nietzsche) were non-ideological, liberal scholars such as Giorel Curran (2006) condemn an earlier generation of anarchists as ideological. They are thus portrayed as being dogmatic, sectarian, doctrinaire—conforming blindly to tradition, authoritarian, vanguardist, confrontational—if not outright violent, and of course "essentialists." In contrast, the "new anarchists"—poetic terrorists, anarcho-primitivists, anarcho-capitalists, Stirnerite individualists, and post-structural anarchists—are described as open, nonviolent, tolerant, flexible and eclectic, concerned with methodology not ideology, and as expressing the "spirit" of anarchism, not some dreadful, misguided anarchist ideology. As with other academics such as John Moore and Ruth Kinna, Curran articulates a rather simplistic, linear, bipolar conception of anarchist history, though Curran admits that there might be some continuity between the "old" and the "new" anarchism.[5]

Three points need to be made regarding this baneful analysis.

The first is that there is no such intellectual or political position or theory that is non-ideological (using the term "ideology" in a non-pejorative sense). For by their thoughts and actions all anarchists and political radicals express an ideology. Autonomist Marxism, anarcho-primitivism, Stirnerite ultra-individualism, postmodern anarchism, and poetic terrorism (ontological anarchy!) are all ideologies, and their devotees are just as ideological as any social or class struggle anarchist.[6]

Of course, the old anarchists were committed to anarchism as a political ideology and movement, whether mutualists (Proudhon), collectivists (Bakunin), anarcho-syndicalists (Rocker), and anarchist communists

(Kropotkin, Reclus, Goldman, Malatesta). All reflected upon, defended and propagated (social) anarchism as a political ideology or tradition. This was thought of as an ongoing developing tradition. But they were not conformists, blindly following a tradition. Kropotkin, for example, was critical of certain aspects of the anarchism of Proudhon and Bakunin, but unlike contemporary academic radicals like Moore, Holloway, and Kinna he did not pretentiously declare them "obsolete," irrelevant, or "outmoded."

Secondly, it would be difficult to find among an earlier generation of anarchists (from Bakunin to Landauer) any that are more doctrinaire, more dogmatic, more sectarian or more "ideological" than the anarcho-primitivists Bob Black and John Moore, given their blanket dismissal of *all* aspects of civilisation, and their equally hostile dismissal of libertarian socialism, and all forms of working class struggle. There is indeed a sense in which anarcho-primitivism is akin to the individualist fringes of American right-wing extremism (Sheehan 2003, 43).

Thirdly, showing no real engagements with the writings and political struggles of the "old" anarchists, Curran's account verges on caricature. For she uncritically embraces all the negative opinions of the detractors of class struggle anarchism—whether liberal academics, ultra-individualists, or postmodernists. And so falsely accuses "old" anarchists of being "ideological," that is lacking the appropriate anarchist temperament or as being devoid of the "spirit" of anarchism. Setting up a radical dichotomy between anarchism as a political tradition, shared by all social anarchists, and the "spirit" of anarchism is quite fallacious. And accusing an earlier generation of anarchists, like Kropotkin, of lacking such a "spirit" is quite unfounded.

Anarchists like Bakunin and Kropotkin—like all radical scholars—were always open to new ideas, and critically absorbed ideas from a wide range of sources—philosophical, cultural, political, scientific. They had far more scholarship than some of their postmodern and liberal critics, but being anarchists and not university academics, they expressed their ideas as lucidly as possible, aiming to reach and appeal to a working class readership. They also gave their ideas some degree of coherence though neither Bakunin, Kropotkin, or Reclus attempted to create some philosophical system in the style of Spinoza or Hegel. All this contrasts markedly with the "new" anarchists, who congratulate themselves on being able to hold incompatible or contradictory premises—advocating for example, both tribal communalism and Stirnerite possessive individualism—as well as expressing their ideas in the most obscurantist, scholastic jargon.

Kropotkin not only embraced and developed (social) anarchism as a political ideology, but in his life and practices expressed also the "spirit" of anarchism—apart from a sad lapse of his principles at the outbreak of the

the First World War. Reclus expressed the "spirit" of anarchism even more (see Morris 2007, Clark and Martin 2013).

9. Conclusions

We can but conclude that there is very little substance in the so-called poststructuralist critique of (social) anarchism. Such a view is certainly confirmed in Richard Day's (2005) excellent study of anarchist currents in recent social movements. For Day acknowledges that Kropotkin long ago formulated an ongoing struggle between the logic of hegemony and the logic of affinity (i.e., anarchism); that his writings on the state "prefigure" those of Deleuze and Guattari in *A Thousand Plateaus* (1988); and that Kropotkin affirmed the creative power of the "masses" not only in resisting oppression but in creating new social institutions through mutual aid and voluntary cooperation. Throughout history this has been achieved by direct action, or what Day describes as "structural renewal." Thus Day concludes that Kropotkin was the "first postanarchist" (2005, 121–23).

Kropotkin, of course, was not "post" anything. But this suggestion does indicate that there is nothing particularly new or original in the politics of the poststructuralists. Indeed, Day continually implies that both the poststructuralist philosophers (Deleuze, Foucault, Derrida) and autonomist Marxists (Hardt and Negri 2005, Holloway 2005) simply appropriated and replicated the basic principles of social or class struggle anarchism, with very little acknowledgement. As he coyly put it: "Deleuze, Foucault or Derrida might owe certain unacknowledged debts to the anarchist tradition" (2005, 94). John Holloway's autonomist Marxist text *Change the World without Taking Power* (2005) simply indicates either a refusal to acknowledge the existence of anarchism or a rather baneful ignorance of this political tradition. For anarchists have been advocating an alternative both to liberal reformism and to the Marxist revolutionary party and workers' state for well over a century. Indeed, ever since William Godwin, who first advocated a communist society without government (Baldwin 1927, 290).[7]

The "logic of affinity," "structural renewal," "changing the world without taking power"—these simply describe the radical practices long advocated by social or class struggle anarchists, practices that have just been rediscovered by the postmodernists!

Notes

1   I am informed that this article is unduly strident and polemical. This polemical tone stems partly from the fact that the essay was initially given as a talk on the "new anarchism" to the Northern Anarchist Network (in September 2007) and partly from my own exasperation at the derisive treatment of an earlier generation of anarchists by many contemporary academics. Needless to say, following a long anarchist tradition

I see no antithesis between polemics and scholarship. But note: although the tone of the paper may be critical and polemical, nowhere do I intend to belittle or misrepresent the work of the scholars I discuss.

2   By social or class struggle anarchism I refer to the kind of anarchism long ago outlined by Bakunin and Kropotkin (see my studies of these anarchists, 1993, 2004). It embraces the following basic tenets: a rejection of the state and all forms of power and authority that inhibit the liberty of the individual; a rejection of capitalism along with its competitive ethos and its possessive individualism (and thus the advocacy of some form of socialism); and, finally, the creation of forms of social organization based on mutual aid and voluntary cooperation, and which enhance and promote the fullest expression of human liberty. Such organizational forms are thus independent of both the state and capitalism (for a contemporary definition of class struggle anarchism see Franks 2006, 12).

3   It is important to understand that this paper is *not* a critique of poststructuralism, anarchist or otherwise. Nor is it an assessment of whether or not poststructuralist (or postmodern) philosophers (or their acolytes) are anarchists. It is, rather, focused specifically on the *misrepresentation* of an earlier generation of class struggle anarchists by the academic devotees of poststructuralism (or postmodernism). My own feeling is that apart from Deleuze and Foucault (at odd moments) none of the poststructuralists (Bourdieu, Baudrillard, Derrida, Lyotard, Rorty, et al.) can be considered anarchists. See my short review of the poststructuralists "The Great Beyond" in *Freedom* 68, no. 23 (2007): 5.

4   For a full discussion of Foucault's complex relationship with the Iranian Revolution see Eribon (1993, 281–91).

5   John Moore's bipolar conception of anarchist history was well expressed in the pages of the *Green Anarchist* (e.g., 57 [1999], 23) where social anarchists are described pejoratively as "leftists," and dismissed as dreary, political racketeers. As well, of course, in his review of Todd May (1994) where Moore explicitly outlines a "two-phase periodisation" of anarchist history (*Anarchist Studies* 5 [1997] 157). He personally suggested to me that Kropotkin's revolutionary anarchism was "obsolete," prompting me to write a book on Kropotkin's anarchism and social ecology! The book (2004) aimed to reaffirm Kropotkin's continuing relevance for anarchist theory.

With regard to Ruth Kinna, although within the chapter "What Is Anarchism" she presents a more nuanced account of anarchism, this completely undermines and runs counter to the key dichotomy she embraces between the "old" and the allegedly "new" anarchism, and her suggestion that "traditional" anarchism (i.e., class struggle anarchism") has now become "outmoded" (2005, 21). Setting up a dichotomy between the "old" and allegedly "new" anarchism is both conceptually and historically highly problematic, if not obfuscating.

6   I use these different radical tendencies not because I oppose them, but because they have been described by Curran (2006) as the "new" radicalism and as being "post ideological."

7   It is of interest that the only mention of anarchism in Holloway's entire text is to declare that anarchism—as distinct from reform and revolution—no longer seems "relevant" to contemporary activists (2005, 21).

References

Albert, Michael. 2006. *Realizing Hope: Life beyond Capitalism*. London: Zed Books.

Baldwin, Roger N. 1927. *Kropotkin's Revolutionary Pamphlets*. New York: Dover Publications.

Barua, Dilip Kumar. 1991. *Edward Carpenter: The Apostle of Freedom*. Burdwan: University of Burdwan.

Bey, Hakim. 1991. *TAZ: the Temporary Autonomous Zone, Ontological Anarchy, Poetic Terrorism*. Brooklyn: Automedia.

Black, Bob. 1997. *Anarchy after Leftism*. Columbia, MO: CAL Press.

Bookchin, Murray. 1980. *Toward an Ecological Society*. Montreal: Black Rose.

_____. 1982. *The Ecology of Freedom*. Palo Alto: Cheshire.

_____. 1995. *Re-enchanting Humanity*. London: Cassell.

_____. 1999. *Anarchism, Marxism and the Future of the Left*. San Francisco: AK Press.

Bronner, Stephen Eric. 2004. *Reclaiming the Enlightenment: Toward a Politics of Radical Engagement*. New York: Columbia University Press.

Brown, L. Susan. 1993. *The Politics of Individualism*. Montreal: Black Rose.

_____. 2004. "Looking Back and Looking Forward: The Radical Humanism of 'The Politics of Individualism.'" *Anarchist Studies* 12, no. 1: 9–15.

Call, Lewis. 1999. "Anarchy in the Matrix." *Anarchist Studies* 7, no. 2: 95–117.

_____. 2003. *Postmodern Anarchism*. Lanham, MD: Lexington Books.

Clark, John P., and Camille Martin, eds. 2013. *Anarchy, Geography, Modernity: The Radical Social Thought of Elisée Reclus*. Oakland: PM Press.

Crowder, George. 1992. "Freedom and Order in Nineteenth-Century Anarchism." *The Raven* 5, no. 4: 342–57.

Curran, Giorel. *21st Century Dissent: Anarchism, Anti-Globalization and Environmentalism*. Basingstoke: Palgrave.

Davis, Angela Y. 2003. *Are Prisons Obsolete?* New York: Seven Stories Press.

Day, Richard J.F. 2005. *Gramsci Is Dead: Anarchist Currents in the Newest Social Movements*. London: Pluto Press.

Deleuze, Gilles, and Félix Guattari. 1988. *A Thousand Plateaus*. London: Athlone Press.

Derrida, Jacques. 1994. *Specters of Marx*. London: Routledge.

Eagleton, Terry. 2003. *After Theory*. London: Penguin Books.

Eribon, Didier. 1993. *Michel Foucault*. London: Faber.

Foucault, Michel. 1980. *Power/Knowledge*. New York: Pantheon Books.

Franks, Benjamin. 2006. *Rebel Alliances*. Oakland: AK Press.

Hardt, Michael, and Antonio Negri. 2005. *Multitude: War and Democracy in an Age of Empire*.

Holloway, John. 2005. *Change the World without Taking Power*. London: Pluto Press.

Kinna, Ruth. 2005. *Anarchism: A Beginner's Guide*. Oxford: Oneworld Publications.

Kropotkin, Peter. 1889. *Words of a Rebel*. Montreal: Black Rose.

_____. 1899. *Fields, Factories and Workshops* (1985). London: Freedom Press.

_____. 1902. *Mutual Aid: A Factor in Evolution*. London: Heinemann.

_____. 1905. *Russian Literature: Ideals and Realities* (1991) Montreal: Black Rose.

_____. 1924. *Ethics: Origins and Development*. Dorchester: Prism Press.

_____. 1970. *Selected Writings on Anarchism and Revolution*. Edited by Martin A. Miller. Cambridge, MA: MIT Press.

_____. 1993. *Fugitive Writings*. Montreal: Black Rose.

_____. 1997. *Small Communal Experiments and Why They Fail*. Petersham NSW: Jura Books.

Lehning, Arthur. 1973. *Bakunin: Selected Writings*. London: Cape.

Lunn, Eugene. 1973. *Prophet of Community: The Romantic Socialism of Gustav Landauer*. Berkeley: University of California Press.

MacPherson, Charles B. 1962. *The Political Theory of Possessive Individualism*. Oxford: Oxford University Press.

Marx, Karl, and Friedrich Engels. 1968. *Selected Works*. London: Lawrence and Wishart.

May, Todd. 1994. *The Political Philosophy of Poststructuralist Anarchism*. Pennsylvania State University Press.

Mayr, Ernst. 2002. *What Evolution Is*. London: Weidenfeld and Nicolson.

Moore, John. 1998. "Maximalist Anarchism/Anarchist Maximalism." *Social Anarchism* 25: 37–40.

_____. 2004. "Attentat Art: Anarchism and Nietzsche's Aesthetics" In *I Am Not a Man, I Am Dynamite: Friedrich Nietzsche and the Anarchist Tradition*, edited by John Moore, 127–40. Brooklyn: Autonomedia.

Morland, David. 2004. "Anti-capitalism and Poststructuralist Anarchism." In *Changing Anarchism: Anarchist Theory and Practice in a Global Age*, edited by Jon Purkis and James Bowen, 23–38. Manchester: Manchester University Press.

Morris, Brian. 1991. *Western Conceptions of the Individual*. Oxford: Berg.

_____. 1993. *Bakunin: The Philosophy of Freedom*. Montreal: Black Rose.

_____. 1994. *Anthropology of the Self*. London: Pluto Press.

_____. 2001. Kropotkin's Metaphysics of Nature. *Anarchist Studies* 9: 165–80.

_____. 2004. *Kropotkin: The Politics of Community*. Amherst, NY: Humanity Books.

_____. 2007. *The Anarchist Geographer: An Introduction to the Life of Peter Kropotkin*. Minehead: Genge Press.

Mumford, Lewis. 1970. *The Pentagon of Power*. London: Secker and Warburg.

Newman, Saul. 2001. *From Bakunin to Lacan*. Oxford: Lexington Books.

_____. 2004. "Anarchism and the Politics of Ressentiment." In *I am Not a Man, I Am Dynamite: Friedrich Nietzsche and the Anarchist Tradition*, edited by John Moore, 107–26. Brooklyn: Autonomedia.

Patton, Paul. 2000. *Deleuze and the Political*. London: Routledge.

Purchase, Graham. 1996. *Evolution and Revolution: An Introduction to the Life and Thought of Kropotkin*. Petersham, NSW: Jura Books.

Purkis, Jon, and James Bowen, eds. 1997. *Twenty-First Century Anarchism: Unorthodox Ideas for the New Millennium*. London: Cassell.

Rabinow, Paul. 1984. *The Foucault Reader*. Harmondsworth: Penguin Books.

Rocker, Rudolf. 1989. *Anarcho-Syndicalism*. London: Pluto Press.

Rose, Hilary, and Steven Rose, eds. 2000. *Alas, Poor Darwin: Arguments against Evolutionary Psychology*. London: Cape.

Rötzer, Florian. 1995. *Conversations with French Philosophers*. Atlantic Highlands, NJ: Humanities Press.

Sheehan, Sean M. 2003. *Anarchism*. London: Reaktion Books.

Ward, Colin. 1973. *Anarchy in Action*. London: Allen and Unwin.

Wieck, David. 1979. "The Negativity of Anarchism." In *Re-inventing Anarchy*, edited by Howard Ehrlich, 138–55. London: Routledge.

Zerzan, John. 1994. *Future Primitive and Other Essays*. Brooklyn: Autonomedia.

13

Ecology and Socialism (2010)

In this essay I aim to explore the relationship between ecology and socialism. But I shall do so in a rather roundabout fashion, making a detour to
discuss political ecology more generally, specifically to outline the varied
political responses to the ecological crisis that now confronts us.

Bookchin and Commoner
Long ago the biologist Paul Sears (1964) described ecology as the "subversive science," and there is no doubt that when I first became involved in
environmental issues in the 1960s ecology was seen very much as a radical
movement. The writings of Murray Bookchin (1971) and Barry Commoner
(1972) emphasised that we were confronting an ecological crisis and that
the roots of this crisis lay firmly with an economic system—capitalism—
that was geared not to human well-being but to the generation of profit,
that saw no limit to industrial progress, no limit to growth and technology,
even celebrating the achievements of what Lewis Mumford (1970) called
the "mega-machine."

Ultimately it was felt, by both Commoner and Bookchin, that capitalism was destructive not only to ourselves but to the whole fabric of
life on the planet. For the underlying ethic of capitalism was indeed the
technological domination of nature, an anthropocentric ethic that viewed
the biosphere as having no intrinsic value; it was simply a resource to be
exploited—by capital.

Over thirty years ago Murray Bookchin was thus describing capitalism as "plundering the earth" in search of profits, and was highlighting
with some prescience—long before Al Gore and George Monbiot—the
problems of global warming, that the growing blanket of carbon dioxide

would lead to destructive storm patterns, and eventually to the melting of the ice caps and rising sea levels (Bookchin 1971, 60)

This was in addition to the many other ecological problems that Commoner and Bookchin identified as constituting a part of the "modern crisis"—deforestation, urbanisation, the impact of industrial farming, pollution of the oceans and atmosphere, toxic chemicals and food additives, and the wanton destruction of wildlife and the subsequent loss of biodiversity.

It is now almost universally recognised that we are indeed beset with many ecological problems—which has of course to be set alongside many other besetting problems: widespread poverty and increasing economic inequalities; the existence of weapons of mass destruction, and what has been described as the "dialectics of violence"; political repression and the denial of basic human rights; and the social and economic devastation that has accompanied the imposition of free-market capitalism.

All these problems are intrinsically inter-linked by what Paul Ekins described as a "single, systematic problematique" (1992, 13)—namely global capitalism.

The Coming Apocalypse?

What is significant about the present ecological crisis is that it is invariably interpreted in terms of a religious metaphysic. Thus environmental problems are described as if we are facing an apocalypse, and we are thus presented with a doomsday scenario; humans are viewed as inherently destructive of the environment, a new rendering of the concept of original sin. Thus humans—at least since the "fall" (the advent of agriculture)—are interpreted as "aliens" or as "parasites" on earth. This view is even expressed by respected biologists such as Edward Wilson (2006, 13). And finally, of course, there is a call for some kind of redemption, and we are therefore all being urged to "save" the planet. Such hubris is quite mind-boggling.

There are, of course, many serious ecological problems we need to address, but there is no call for scare-mongering or apocalyptic visions.

Darwin's Evolutionary Ecology

Ecology, and the development of an ecological worldview and an ecological sensibility, has, of course, a long history. It emerged essentially towards the end of the nineteenth century, and the key figure in this development was the naturalist Charles Darwin. Although it is of interest to note that several of the gurus of the ecology movement hardly even mention Darwin (e.g., Naess 1989, Shiva 1989, McKibben 1990, Merchant 1992). For what was significant about Darwin's evolutionary ecology was that it established the following:

- It introduced the idea that humans were not the special products of god's creation but evolved according to principles that operated throughout the natural world: that we are simply a third chimpanzee.
- It stressed that humans were an intrinsic part of nature, and thus undermined completely—long before quantum physics (Capra), phenomenology (Heidegger) and feminist philosophy (Plumwood)
- It undermined the mechanistic worldview, along with its Cartesian dualisms, its essentialism, and its anthropocentric ethic.
- It emphasised the crucial importance of openness, creativity, chance, probability, and the agency and individuality of all organisms in the evolutionary process; and finally,
- It suggested a way of understanding that was Naturalistic and historical, not spiritual or theological, thus dispensing with all notions relating either to some divinity or to various spiritual beings.

Early Eco-Socialism

At the end of the nineteenth century, during the first wave of the ecology movement—which was in fact an early response to the environmental impact of industrial capitalism—there seems to have been a close relationship between socialism and a "green" sensibility, as Peter Gould emphasised in his book *Early Green Politics* (1988). Many of the early socialists were involved in ecological issues and attempted to combine socialism, with its emphasis on social justice and the class struggle with an ecological sensibility involving a respect for nature and an ecological critique of industrial capitalism. William Morris, Robert Blatchford, Peter Kropotkin, Henry Salt, and Edward Carpenter were all, in their different ways, eco-socialists.

History of Ecology

But there is no intrinsic connection between socialism and ecology: to the contrary, many of the pioneers of ecology embraced extreme right-wing politics. As Anna Bramwell (1989) indicated in her history of the ecology movement—reflected in the advocacy of wildlife conservation, organic farming, and the back-to-the-land movement—many of the early exemplars of an ecological perspective were fascists, or at least had right-wing sympathies. They include Rolf Gardiner, Henry Williamson, and John Hargrave, as well as some of the iconic figures in the German National Socialist movement such as Walther Darre, Rudolf Hess, and Martin Heidegger. Thus one has to acknowledge that an ecological ethic or sensibility has been embraced by people right across the political spectrum.

It may therefore be useful to briefly outline some of the contrasting political responses to the ecological crisis that have been described in the literature on political ecology. I shall delineate seven key responses.

1. Authoritarian Ecological Politics

Although there is indeed a class association between ecology and fascism, leading many socialists to dismiss the environmental movement itself as inherently reactionary, what is significant is that many of the responses to the ecological crisis do in fact entail an extremely authoritarian politics. Given the alleged gravity of the situation, especially in relation to population growth and resource depletion, many well-known scholars have thus suggested that the only solution to the crisis entails the creation of a strong central authority. In the words of William Ophuls (1977), the choice we have is between "Leviathan or oblivion."

Thus Garret Hardin, well known for his essays on the "Tragedy of the Commons"—which is essentially a plug for free-market capitalism and indicates no understanding of the concept of the "commons"—advocates a geopolitics in which the state has draconian powers to regulate reproduction and curtail environmental degradation. Small wonder that Hardin has been described as an "eco-fascist" (Hay 2002, 174–77).

Likewise Rudolf Bahro. Early in his career Bahro was a leading party intellectual of the East German Communist Party, and wrote a well-known critique of Soviet communism, *The Alternative in Eastern Europe* (1978). He then became a "visionary green theorist" and leader of the "fundamentalist" wing (fundis) of the West German Green Party. At the end of his life, however, he became a religious fundamentalist—of a mystical gnostic variety—and a strong advocate of an authoritarian "salvation government," even making a plea for a "Green Adolf." It is hardly necessary to note that contemporary neo-Nazis explicitly see themselves as radical environmentalists.

2. Deep Ecology

Towards the end of last century many people in the ecology movement embraced a current of thought that was known as "deep ecology." The term was coined by the Norwegian philosopher Arne Naess, and "deep ecology" was explicitly seen as a radical movement, in contrast to the "shallow" environmentalism of those merely advocating reformist measures.

Deep ecology was characterised by the following: a neo-Malthusian tendency which, following the likes of Paul Ehrlich and Garret Hardin, views human population growth as a primary factor in an impending ecological catastrophe; an emphasis on the preservation of wilderness areas, to the complete neglect of agriculture and the urban context; a tendency to invoke a new kind of "original sin" in which an undifferentiated humanity is seen as a destructive agent that threatens the very survival of life on earth; an advocacy of biocentrism, or "bio-spherical egalitarianism," an ethic that invariably tended to denigrate into misanthropic or anti-humanist sentiments—deep ecologists, for example, applauding the AIDS epidemic as a way of reducing the human population; and finally, a tendency to view the ecological crisis in essentially ideological terms—as due to the pervasive influence of anthropocentrism, the ethic that suggests that humans are separate from, and superior to the rest of nature, and that this justifies the use of nature simply as a resource. In contrast, deep ecologists advocate an "ecological consciousness" that allegedly has its sources and exemplars in a bizarre and motley collection of philosophies and religious traditions (animism) of tribal people; Eastern spiritual traditions such as Taoism and Buddhism; the Christian tradition of St. Francis; Gandhi's advocacy of Advaita Vendanta; the philosophies of Spinoza, Whitehead, and Heidegger; and the "new physics" as interpreted by Fritjof Capra (Devall and Sessons 1985).

Murray Bookchin has made some trenchant criticisms of deep ecology, not only for its anti-humanism and neo-Malthusian tendencies, but for completely ignoring capitalism and the social origins of the ecological crisis. Although deep ecologists like Naess (1989) suggest that being "green" transcends the opposition between "blue" (conservatism) and "red" (socialism—equated with state capitalism), in fact the politics of most eco-ecologists seem to envisage the continued existence of both capitalist corporations and the nation-state.

For example, deep ecologists like Edward Abbey and Dave Foreman—both avowed neo-Malthusians—deem the United States government to be essential in order to curb further immigration—especially across the Mexican border. For deep ecologists like Robin Eckersley the state continues to be extolled as the agent for economic planning and the market (capitalism) is acknowledged as the basic arbiter to resource allocation. All this is little different from the mixed economy of the liberal democrats, politically it is essentially liberal and reformist (Naess 1989, Eckersley 1992).

3. Spiritual Ecology

Closely linked to, and often identified with, deep ecology, is spiritual ecology, perhaps the most pervasive response to the ecological crisis. In fact, as soon as one begins to discuss ecology one is invariably faced with a false dilemma. Either we have to side with religious mystics and neo-pagans and activate a "sacramental" or religious attitude to nature, or we are alleged to align ourselves with aggressive imperialism and industrial capitalism. The stark choice we are given is either between mechanistic philosophy and the technological mastery of nature or embracing metaphysics; it is a (false) choice between either mammon (industrial capitalism) or god (religion).

A typical example is David Watson's book *Against the Megamachine* (1999), where the choice we are given is either the "prison house of urban industrial civilization" with its accompanying ideologies, or "primitivism"—entailing the wholesale rejection of technology, the affirmation of a hunter-gatherer existence, and the embrace of neopaganism—tribal animism.

In fact almost every religion under the sun is now embracing an ecological sensibility, and is being, therefore, an appropriate response to the ecological crisis. Sometimes a religious metaphysic is combined with a libertarian politics as with Starhawk's (1982) embrace of the mother-goddess religion, and Peter Marshall's (1998) extolling of new age religion—an eclectic embrace of Taoism, Buddhism, and Advaita Vedanta. Although heralded as a "new philosophy" of nature, it simply indicates a reaffirmation of ancient religious traditions.

What is ignored by all the spiritual ecologists is that there is another ecophilosophical tradition, stemming from Darwin and the early literary naturalists, that repudiates both mechanistic philosophy, along with its ethnocentrism and the ethic implying the "conquest" of nature and spiritualism, or any form of religious metaphysic. It is in fact an ecological worldview that is quite distinct from any form of spiritualism.

4. Eco-Primitivism

While spiritual ecologists view the present crisis as a spiritual crisis, and tend to repudiate the radical elements of the eighteenth-century Enlightenment, the eco-primitivists—such as Watson—view the primary cause of the present crisis as civilisation itself. The key figures in this tendency to see "civilisation"—interpreted as "Leviathan" or the "Beast"—as the prime factor in the destruction of both human freedom and the environment, are Fredy Perlman (1983) and John Zerzan (1994).

Both tend to view our hunter-gatherer past as an idyllic era of virtue, freedom and authentic living—a reaffirmation of the myth of the "noble savage." Thus the last thousand years or so of human history after the advent of agriculture (the fall) is seen as a period of tyranny and hierarchical control, and as involving a loss of contact with the world of nature and thus the continual degradation of the environment. All those products of the human imagination—farming, art, literature, philosophy, technology, science, urban living, symbolic culture—tend to be reviewed negatively by Zerzan in a monolithic sense. The future, he tells us, is primitive.

Although eco-primitivists continually plead that they are not advocating a return to the stone-age, nevertheless to repudiate entirely urban living, technology, and agriculture in a world that currently sustains around six billion people hardly makes sense. Thus it is not surprising that eco-primitivism has been widely critiqued by radical ecologists (Bookchin 1995; Albert 2006, 178–84) as well as by neoliberals (Whelan 1999).

5. Green Capitalism

At a meeting of deep ecologists in 1991 to discuss Earth Day 1990 and the present ecological crisis, many of the participants not only made an appeal to governments to deal with ecological problems, but advocated—as a response—the privatisation of land and wildlife (Oelschlaeger 1992). It was thus felt that contemporary environmental problems could best be addressed by an appeal to big business, by extending private property rights and using market mechanisms. Capitalism was thus seen not as the cause of environmental problems but as the solution.

Green capitalism was therefore embraced not only by some deep ecologists, but by a plethora of neoliberals, and this coincided with the emergence of the concept of "sustainable development." Given that the capitalist system is based on the logic of capital accumulation, and the maxim "grow or die," sustainable development appears to be something of an oxymoron. But what it essentially extols is sustainable growth (capitalism).

Yet, surprisingly—or not so surprisingly!—over the last three decades numerous books have been published that promote green capitalism or free-market environmentalism (for example Hawken 1993). It takes various forms, but it is initially worth noting that some of the most ardent environmentalists are in fact capitalists. Al Gore, Zac Goldsmith (owner of *The Ecologist* magazine) and Lord Peter Melchert—well-known for his campaign against genetically modifying crops—are all millionaires, while Jonathon Porritt, ex-guru of the Green Party and now adviser to New Labour on ecological issues, has recently written a book extolling the virtues of capitalism (2005).

It is also worth noting that many ecological foundations, such as Sara Parkin and Jonathon Porritt's "Forum of the Future" are financially supported by such multinational corporations as ICI, BP, and Tesco (Heartfield 2008, 31). With regard to contemporary green capitalism three tendencies are worth noting.

One is that capitalist corporations are now in the process of "greening" their public image, often supported by leading green activists. This is something that the Shell corporation has been engaged in for many decades, given its awful record in terms of environmental destruction. It would be difficult to find any major multinational corporation these days that does not proudly proclaim and advertise its ecological sensibility and its green credentials. Shell, Rio Tinto, McDonalds . . . all pronounce a commitment to "sustainable development."

Secondly, although most people now acknowledge that there is an environmental crisis, efforts are continually being made to convince us that this is due to a lack of spirituality, or that there are too many people on earth, or for example that deforestation is caused by poor peasants—not by logging companies, mining corporations such as Vedanta and Rio Tinto, and expanding ranching enterprises that cater for the increasing demand for meat. But what equally clouds the issue is the suggestion that global warming and other environmental issues have nothing to do with an economic system geared to growth and profit: it is solely due to the actions of individual "consumers." So we are all being urged to do what we can to "save" the planet.

Finally, this laudable concern for the environment on the part of transnational corporations is clearly a front to enable such corporations to seek further opportunities for capitalist expansion and for generating even more profit. Thus, for example, the expansion and export of the nuclear industry to all parts of the world is now being heralded as a great way of cutting "carbon emissions" and thus helping to "save" the planet. But at what ecological cost?

With Jonathon Porritt and many other environmentalists the answer to the present ecological crisis is thus a marriage of capitalism and spiritualism.

6. Eco-Marxism

When environmental issues came to the forefront in the 1970s Marxists tended to be rather sceptical of, or even hostile to, the ecology movement. It was felt to be a middle class movement, with a strong neo-Malthusian orientation, and that it tended to divert attention away from the class struggle, and from such issues as world poverty, imperialism, racism, and social justice. Those in the ecology movement were dismissed as "bourgeois" and as "ecofreaks." Marxism and ecology were seen as almost incompatible with worldviews, as they still are by writers like James Heartfield, who describes "environmentalism" as in essence "the ideology of capitalism" (2008, 91).

At the same time within the green movement Marx, and Marxism more generally, was almost universally interpreted as lacking any ecological sensibility, as being, in fact, anti-ecological. This was well expressed by John Clark (1984), who argued that Marx's project of human emancipation was essentially based on two principles: economic determinism and a "productivity" view of human nature which viewed human freedom as involving the "domination" or "mastery" of nature. This, he felt, sidelined issues relating to culture and communication. Even in its best moments, Clark wrote, Marx's philosophy remained "largely uncritical of the industrial system of technology and the project of human domination of nature" (1984, 27).

During the last two decades of the twentieth century two things happened. One was that there was a determined effort to forge a kind of alliance between the labour movement, and socialism more generally, and the green movement, as it was then described. There was in fact a concentrated effort at "greening of Marxism" (Benton 1996). Such an integration of Marxism (socialism) and ecology took many forms, but was reflected in such books as Joe Weston's edited collection *Red and Green: The New Politics of the Environment* (1986) and David Pepper's *Ecosocialism* (1993), as well as in more problematic contributions by André Gorz (1980) and Martin Ryle (1988). Gorz advocated a mish-mash of quite incompatible strategies; high tech and human-scale "convivial tools"; local control and state planning; socialism and private enterprise. Hardly surprising that many felt that Gorz lacked any ecological sensibility and could hardly be described as a Marxist. On the other hand Ryle's *Ecology and Socialism*, like the manifesto of the Green Party, simply advocated a form of market socialism, based on petty commodity production—as suggested by Proudhon over a hundred years ago!—which both Marx and Kropotkin had critiqued long ago.

The other measure was that a determined effort was made to refute the idea that Marx (and Engels) can be understood as simply worshippers of productive or as advocates of the mechanistic domination of nature. The key figure here is John Bellamy Foster (2000) who sought to reaffirm Marx's ecological credentials. Foster emphasised the following: that Marx embraced a worldview that was not only materialist but was deeply ecological; that ecology was central to Marx's thinking; that Marx, following Darwin, completely repudiated the mechanistic materialism of the eighteenth century, and thus stressed that humans were an intrinsic part of nature; that anthropocentrism, and the radical separation of humans from nature, was essentially a product of capitalism, with its separation of capital and labour; that Marx was aware and critical of the ecological problems relating to large-scale industry and capitalist agriculture, especially in relation to soil conservation (capitalist production, Marx wrote, undermines "the original sources of all wealth—the soil and the worker," *Capital* 1, 548); and finally, that although Marx put an important emphasis on production and labour—as the essential interaction between humans and nature—Marx made a crucial distinction between production for profit (exchange) and production for use, and envisaged that production in a future socialist society would entail both the social ownership of the means of production and a free association of producers.

Central to eco-Marxism is the idea that the main course of the ecological crisis is the capitalist mode of production. Capitalism is, as Joel Kovel (2002) described it in his eco-socialist manifesto, "the enemy of nature." Kovel, in fact, provides us with a powerful ecological critique of capitalism. Unfortunately, Kovel continues to advocate a market economy and statist politics; envisages an unholy marriage between Marxism and spiritualism (mysticism); and seems to lack any real understanding of anarchism—libertarian socialism.

7. Social Ecology

Of all political traditions anarchism has certainly had the worst press. It has been ignored, maligned, ridiculed, abused, misunderstood and misrepresented by writers from all sides of the political spectrum. Theodore Roosevelt described anarchism as a "crime against the whole human race." There are, of course, many varieties of anarchism: there are mutualists, collectivists, Stirnerite individualists, radical aesthetes who embrace Nietzsche (otherwise known as poetic terrorists), anarchosyndicalists, reformist anarchists, eco-primitivists, postmodern anarchists (the latest brand), religious anarchists like Tolstoy and Berdyaer, and anarcho-communists. What they have in common is an emphasis on the freedom and autonomy of the individual, and thus the repudiation not only of the state but of all forms of coercive authority.

But the central tradition has been anarcho-communism, which emerged towards the end of the nineteenth century. It is associated with such figures as Peter Kropotkin, Errico Malatesta, Emma Goldman, Alexander Berkman, and Rudolf Rocker. This form of anarchism is synonymous with libertarian socialism—indeed Kropotkin defined anarchism as revolutionary socialism—and the tendency to see some kind of opposition between anarchism and socialism is quite misleading—both historically and politically. Anarcho-communism is a form of socialism.

Unlike Marxists generally, many anarchists have been engaged with ecological issues, particularly agrarian problems, and social ecology is a current thought that stems from Elisée Reclus and Peter Kropotkin and combines an ecological sensibility and worldview with anarchism or libertarian socialism. The key figures in this form of ecological anarchism besides Reclus and Kropotkin, are Murray Bookchin (1982), Graham Purchase (1997), and the present writer (Morris 1996, 2004). We can outline the main ideas of social ecology in summary fashion.

Like Erich Fromm, social ecologists stress that there is an essential "paradox" at the heart of human life, that there is an inherent duality in social existence; for on the one hand, humans are an intrinsic part of nature, they are natural beings. But, on the other hand, through conscious experience and human culture, we are, in a sense, separate from nature. Following Cicero, Bookchin refers to this human symbolic culture as our "second nature." We are then also intrinsically social and cultural beings.

Fully embracing Darwin's evolutionary theory, social ecologists, as with Marx, emphasise that humans are the product of natural evolution. There is therefore no radical dichotomy between humans and the natural world. This implies a repudiation of Cartesian philosophy—its mechanistic paradigm, its anthropocentric ethic, which envisages the technological mastery of nature, and its atomistic epistemology.

Following Darwin, social ecologists also emphasise that the world—the earth—is neither a spiritual cosmos nor a lifeless machine, but an evolutionary process, which can only be understood by an organic, developmental way of thinking—dialectically. They thus stress the importance of historical understanding, especially in relation to biology. Social ecologists are therefore committed evolutionary naturalists.

This means they are equally critical of much social science, which emphasises a radical dichotomy between culture and nature—even oblating biology completely—and the kind of reductive materialism which tends to downplay the uniqueness of the human species—our "humanness." Thus Bookchin (1995) has made some strident criticisms of neo-Malthusian doctrines, postmodernism, Social Darwinism, and sociobiology.

Remaining true to the radical aspects of the Enlightenment tradition, social ecologists emphasise the need to uphold its fundamental values; namely, liberty and the freedom of the individual, equality and social justice, fraternity and social solidarity, and the Enlightenment's cosmopolitan ethic with its religious tolerance.

Reacting against the Social Darwinian emphasis on conflict, struggle, and the "survival of the fittest," as well as against the atomism inherent in mechanistic science, social ecologists stress the importance of mutual aid and symbiosis in understanding the biosphere, as well as of human life. They therefore embrace and develop Kropotkin's ecological vision, and are critical of the Baconian ethic that envisages the human domination of nature and the treatment of the natural world simply as a resource.

What is important then is that social ecologists, like Marx, embrace an ecological worldview that essentially stems from Darwin. And long before most Marxists, Bookchin, for example, was highlighting the degradation of the natural world under industrial capitalism, and suggesting that the ecological crisis has its "roots" in an ever expanding industrial capitalism, obsessed with economic growth and competition, a market economy that was geared to profits and power, rather than to human needs. To interpret Bookchin as either anti-Marx (Kovel) or as a reformist is to misunderstand Bookchin's libertarian socialism—which is radically anti-capitalist.

The measures needed to overcome the ecological crisis, suggested by social ecologists—from Kropotkin to Purchase—include the following: the decentralisation of the social economy and the development of an economic system based on social ownership and voluntary associations; the integration of the city and the countryside to form "bioregional" zones, thus putting an end to the urbanisation of the landscape; the establishment of participatory forms of democracy, involving local assemblies and direct democracy—what Bookchin termed "libertarian municipalism"—linked to some form of federalism; and finally, the scaling down of technology to a "human scale."

It is important to recognise that social ecologists are not neo-romantics; in fact they are critically engaged and reaffirmed the radical aspects of the Enlightenment tradition, and were neither anti-science nor anti-technology. Although critical of many aspects of modern science and technology—especially their symbiotic relationship with capitalism—social ecologists affirmed the importance of science and of an ecologically informed technology. In contrast with eco-primitivists and some deep ecologists, Bookchin and the social ecologists affirmed the importance of city life and refused to make a "fetish" out of the "wilderness" concept. But they recognised that important lessons could be drawn from studying early tribal societies—the absence of coercive and domineering values; a sense of communal property and the emphasis on mutual aid and usufruct rights; and their expression of an ecological sensibility—which is not to be equated with any religious metaphysic.

The social ecologists also emphasised the positive and creative aspects of human interactions with nature, and the importance of cultural landscapes—which actually constitute the living environment of most humans. What they insisted upon was the need for diversity, and thus the need to develop and conserve wilderness areas (natural landscapes), the countryside (cultured landscapes such as woods, parks, meadows, gardens, and cultivated fields) and urban settings, the town or city duly scaled to human needs and human well-being.

As libertarian socialists, the social ecologists emphasised the importance of people taking control of their own lives, seeking to undermine both capitalism and the state, along with all forms of social oppression, as well as attempting to create an alternative society based on mutual aid and voluntary associations, social solidarity and workers' self-management. Social ecologists thus see no conflict between the struggle for social justice and the struggle to create an environment that is conducive to the well-being not only of humans but of all forms of life.

But as anarchists, social ecologists, unlike eco-Marxists such as Kovel, put little faith either in the parliamentary road to socialism (reformism) or in a vanguard party and the "dictatorship of the proletariat" (Bolshevism), for as anarchists have always stressed, socialism without liberty only leads to tyranny and state capitalism. And socialism without an ecological sensibility is also barren.

References

Albert, Michael. 2006. *Realizing Hope*. London: Zed Books.
Bahro, Rudolf. 1978. *The Alternative in Eastern Europe*. London: New Loft Books.
Benton, Ted, ed. 1996. *The Greening of Marxism*. New York: Guilford Press.
Bookchin, Murray. 1971. *Post-Scarcity Anarchism*. London: Wildwood House.

_____. 1982. *The Ecology of Freedom*. Palo Alto: Cheshire Books.

_____. 1995. *Re-enchanting Humanity*. London: Cassell.

Bramwell, Anna. 1989. *Ecology in the 20th Century*. New Haven: Yale University Press.

Clark, John. 1984. *The Anarchist Moment*. Montreal: Black Rose.

Commoner, Barry. 1972. *The Closing Circle*. London: Cape.

Devall, Bill, and George Sessions 1985. *Deep Ecology*. Salt Lake City: Peregrine Smith.

Eckersley, Robin. 1992. *Environmentalism and Political Theory*. London: UCL Press.

Ekins, Paul. 1992. *A New World Order*. London: Routledge.

Foster, John Bellamy. 2000. *Marx's Ecology: Materialism and Nature*. New York: Monthly Review Press.

Gorz, André. 1980. *Ecology as Politics*. Boston: South End Press.

Gould, Peter C. 1988. *Early Green Politics*. Brighton: Harvester Press.

Hawken, Paul. 1993. *The Ecology of Commerce*. New York: Harper Collins.

Hay, Peter. 2002. *A Companion to Environmental Thought*. Edinburgh: Edinburgh University Press.

Heartfield, James. 2008. *Green Capitalism: Manufacturing Scarcity in an Age of Abundance*. London: Openmute.

Kovel, Joel. 2002. *The Enemy of Nature*. London: Zed Books.

Marshall. Peter. 1998. *Riding the Wind: An New Philosophy for a New Era*. London: Cassell.

McKibben, Bill. 1990. *The End of Nature*. London: Penguin.

Merchant, Carolyn. 1992. *Radical Ecology*. New York: Routledge.

Morris, Brian. 1996. *Ecology and Anarchism: Essays and Reviews on Contemporary Thought*. Malvern Wells: Images.

_____. 2004. *Kropotkin: The Politics of Community*. Amherst, NY: Humanity Books.

Mumford, Lewis. 1970. *The Pentagon of Power*. London: Secker and Warburg.

Naess, Arne. 1989. *Ecology, Community and Lifestyle*. Cambridge University Press.

Oelschlaeger, Max, ed. 1992. *After Earth Day: Continuing the Conservation Effort*. Denton: University of Texas Press.

Ophuls, William. 1977. *Ecology and the Politics of Scarcity*. San Francisco: Freeman.

Pepper, David. 1993. *Ecosocialism: From Deep Ecology to Social Justice*. London: Routledge.

Perlman, Fredy. 1983. *Against His-Story, Against Leviathan!* Detroit: Red and Black.

Porritt, Jonathon. 2005. *Capitalism as If the World Matters*. London: Earthscan.

Purchase, Graham. 1997. *Anarchism and Ecology*. Montreal: Black Rose Books.

Ryle, Martin. 1988. *Ecology and Socialism*. London: Century Hutchinson.

Sears, Paul. 1964. "Ecology: A Subversive Science." *Bioscience* 14: 11–13.

Shiva, Vandana. 1989. *Staying Alive*. London: Zed Books.

Starhawk. 1982. *Dreaming the Dark*. London: Unwin Hyman.

Watson, David. 1999. *Against the Megamachine*. Brooklyn: Autonomedia.

Weston, Joe, ed. 1986. *Red and Green: The New Politics of the Environment*. London: Pluto Press.

Whelan, Robert. 1999. *Wild in Woods: The Myth of the Noble Savage*. London: Institute of Economic Affairs.

Wilson, Edward O. 2006. *The Creation: An Appeal to Save Life on Earth*. New York: Norton.

Zerzan, John. 1994. *Future Primitive and Other Essays*. Brooklyn: Autonomedia.

14

The Revolutionary Socialism of Peter Kropotkin (2010)

1. Introduction

In the opening pages of my book on Michael Bakunin (1993) I offered a quote from the Ghanaian writer Ayi Kwei Armah. It reads: "The present is where we get lost—if we forget our past and have no vision of the future."

Drawing on the past does not entail that we engage in a kind of ancestor worship, any more than envisioning a better future for humankind entails that we become lost in utopian dreams. Nobody chides biologists for having an interest in the work and theories of Charles Darwin, nor should socialists feel embarrassed in examining and drawing on the work of an earlier generation of socialist theorists—not as historical curiosities, but as a source of inspiration and ideas. In this article I want to critically explore the writings of Peter Kropotkin (1842–1921), focusing on his politics and his critique of the Marxist theory of the state as an agency of revolutionary transformation. The writing of the essay has been provoked by the numerous self-styled anarchists—though they are invariably Stirnerite individualists, anarcho-primitivists, or anarcho-capitalists—who join forces with Marxists and liberals in declaring that the ideas of Bakunin and Kropotkin are "obsolete" or have no relevance to present-day political struggles. In fact, anarchism, revolutionary socialism, is the only tenable political alternative to neoliberalism.

2. Kropotkin and Anarchism

As a political philosophy, anarchism has had perhaps the worst press. It has been ignored, maligned, ridiculed, abused, misunderstood, and misrepresented by writers from all sides of the political spectrum: liberals, Marxists, democrats, and conservatives. Theodore Roosevelt, the American president, famously described anarchism as a "crime against the whole human race" (Tuchman 1966, 71), and in common parlance anarchy is invariably linked with disorder, violence, and nihilism. A clear understanding of anarchism is further inhibited by the fact that the term "anarchist" is applied to a wide variety of different philosophies and individuals, as can be seen from Peter Marshall's (1992) well-known history of anarchism. Thus Gandhi, Spencer, Tolstoy, Berdyaev, Stirner, Ayn Rand, Nietzsche, along with more familiar figures such as Proudhon, Bakunin, and Kropotkin, have all been described as anarchists. This has enabled liberal and Marxist scholars to dismiss "anarchism" as a completely incoherent philosophy. It isn't. For what has to be recognised is that anarchism is fundamentally a historical movement and political tradition that emerged only around 1870, mainly among the working class members of the International Working Men's Association, widely known as the First International. Although they did not initially describe themselves as anarchists but rather as "federalists" or as "anti-authoritarian socialists," this group of workers adopted the label of their Marxist opponents and came to describe themselves as "Anarchist Communists." Anarchism as a political movement and tradition thus emerged among the workers of Spain, France, Italy, and Switzerland in the aftermath of the Paris Commune, and among its more well-known proponents were Elisée Reclus, François Dumartheray, Errico Malatesta, Carlo Cafiero, Jean Grave, and Peter Kropotkin.[1]

Although Kropotkin himself described "anarchist communism" as the main current of anarchism, it is in fact virtually synonymous with anarchism as a historical movement, which between 1870 and 1930 spread throughout the world and was thus by no means restricted to Europe. Anarchism, anarchist communism, and libertarian or revolutionary socialism can therefore be regarded as synonyms.[2]

The main inspiration of this movement was Michael Bakunin, who also, as I tried to emphasise elsewhere (1993), was an important political theorist in his own right, for it was Bakunin who first articulated in a coherent fashion a theory of anarchism as libertarian socialism. Importantly, contemporary scholarship has indicated that Bakunin was by no means the "intellectual buffoon" bent on violence, destruction, and millennial dreams, as his liberal and Marxist detractors have tended—quite falsely—to portray him.[3]

But the key figure in the development of anarchist communism as a coherent political tradition was Peter Kropotkin, who towards the end of the nineteenth century wrote a series of essays and tracts outlining the basis and principles of anarchist communism. These were later published as pamphlets, or in book form, two texts being particularly significant—*Words of a Rebel* (1885) and *The Conquest of Bread* (1892).

Kropotkin came from the upper echelons of the Russian aristocracy and was in fact born a prince. After military service in Siberia and Manchuria, where he spent much of his time undertaking geographical research, Kropotkin soon became involved in radical politics and at the age of thirty he joined the Chaikovsky Circle of revolutionary populists in St. Petersburg. What prompted this was that in the spring of 1872 Kropotkin had spent some time in Switzerland. There he met members of the Jura Federation, many of whom were friends of Bakunin, joined the International Working Men's Association, and came to have discussions with some of the leading exiled communards, such as Louise Michel, Elisée Reclus, and Benoît Malon. Surprisingly, he never met Bakunin. But his experiences in Switzerland as he recorded in his *Memoirs* (1899, 287), led Kropotkin to become a committed libertarian socialist, an anarchist. Imprisoned twice for his revolutionary activities and spending much time in France, Kropotkin eventually settled in Britain in 1886. He was to remain in Britain as an exile for over thirty years. In exile he seemed to combine the roles of a socialist prophet and scientist, and earned a living as a scientific journalist. An accomplished geographer, Kropotkin was a pioneer ecological thinker. His portrait still hangs in the headquarters of the Royal Geographical Society in London. Kropotkin eventually returned to Russia in 1917, having alienated all his anarchist friends by supporting the Allies in the First World War. He died in 1921 having just completed a philosophical study of *Ethics*. He was buried in Moscow, and his funeral procession, as Victor Serge records, was the last great demonstration against the Bolshevik tyranny (1963, 124).[4]

3. Anarchist Communism

Anarchist communism or revolutionary socialism was established as a political doctrine largely as a reaction against both the economic theories of Pierre-Joseph Proudhon, who was the first to positively describe himself as an anarchist, and the authoritarian politics of the revolutionary communist Karl Marx. Thus anarchist communists envisage a society in which, as Kropotkin puts it, "all the mutual relations are regulated, not by laws, not by authorities, whether self-imposed or elected, but by mutual agreements between members of that society" (Baldwin 1970, 157).

Kropotkin's basic premise was thus a deep and fundamental commitment to individual freedom, and to the self-development of the human person. Through mutual agreements and free association what Kropotkin sought was the "most complete development of individuality combined with the highest development of voluntary association in all its aspects . . . for all imaginable aims" (123).

But for Kropotkin such freedom could only be exercised within a social context, and in a free society, not in a society based on hierarchy and exploitation. It is futile, he wrote, "to speak of liberty as long as economic slavery exists" (124).

Seeing anarchist communism as a synthesis of radical liberalism and communism, Kropotkin repudiated both the state and capitalism. He thus saw a future socialist society as implying the emancipation of humans from both the powers of capitalism and of coercive government. This implied the rejection of all forms of government, including representative government, as well as of the market economy, and what Kropotkin described as the "wage system." Production would be achieved through voluntary associations and self-management, geared to human need and not for profit, and distribution would follow the old adage: "From each according to his ability, to each according to his needs." Land and all other means of production would be held in common, not as private property.

Thus for Kropotkin mutual aid, social solidarity, individual liberty through free cooperation was to be the basis of social life, and he came to describe anarchism as the "no-government system of socialism" (46). The functions of government were to be replaced by local communities and local assemblies, united through a federal system, and Kropotkin insisted on the importance of maintaining and enlarging what later scholars have described as the human life (*Lebenswelt*), "the precious kernel of social customs without which no human society can exist" (137).

In their important study *Black Flame* (2009) Van Der Walt and Schmidt argue strongly that anarchism is a form of libertarian socialism, and for historical reasons should be identified with anarchist communism. They thus repudiate the idea that Godwin, Tolstoy, Proudhon, Stirner, and Tucker are, in fact, anarchists, even though they may have expressed libertarian sentiments. Kropotkin, however, is quite willing to admit that they are anarchists; he simply affirms that anarchism is best understood as libertarian socialism. The tendency of Marxists and Stirnerite individualists to make a radical dichotomy between anarchism and socialism is therefore, on both conceptual and historical grounds, quite misleading, and distorts our understanding of socialism.

4. Kropotkin and State Socialism

Throughout his life Kropotkin was critical of three important radical traditions of the nineteenth century: mutualism, individualist anarchism and what is now described as Marxism, but which Kropotkin described as "state socialism," or implied that it involved "centralized state capitalism" (Baldwin 1970, 185–86), or he simply described the Marxists as "our Social-Democratic friends" (1988, 33).

Although Kropotkin acknowledged that Proudhon expressed libertarian sentiments, and advocated a society without a state—and used the term "anarchy" to describe it—Kropotkin was always critical of the radical tradition that became known as mutualism. This tradition was also embraced by many American individualist anarchists, prominent among them being Josiah Warren, Lysander Spooner, and Benjamin Tucker. What Kropotkin objected to with regard to mutualism was its stress on egoism and the right of the individuals to suppress others if they have the power to do so, its affirmation of private property, petty commodity production, and the wage system (thus, a market economy) and in justifying the use of violence to enforce agreements and defend private property.[5]

Kropotkin was equally critical of the kind of individualism (egoism) expressed by Max Stirner, suggesting that it was a metaphysical doctrine remote from real social life, and bordered on nihilism. He stressed that it was meaningless to emphasise the supremacy of the "unique one" or the development of the human personality in conditions of oppression and economic exploitation. It is of interest to compare Kropotkin's considered and respectful critique of Stirner's work, with that of Engels who dismissed it with such epithets as "arrogant banality," "abysmal triviality," and "bombastic drivel" (Kolpinsky 1972, 30).

Being a member of the First International, Kropotkin had a fairly clear understanding of the kind of politics expressed by Marx and Engels, and he was certainly right to describe Marxism as a form of state socialism. Even though for Kropotkin and other anarchists state socialism (like market socialism) was a contradiction in terms and was perhaps better described as "state capitalism."

In his well known essay "The State: Its Historical Role" (1896), Kropotkin emphasised, like Marx, the intrinsic symbiotic relationship between state power and capitalism, whether this implied laissez-faire, welfare or state capitalism. Though of comparatively recent origin, throughout history, the essential function of the state, Kropotkin argued, had been to uphold systems of hierarchy, and class exploitation, and the modern nation-state, representative government was no different. For Kropotkin "parliamentary rule was capital rule" (1988, 41). In fact, without the machinery of government, Kropotkin suggested, the capitalist system would collapse overnight. The state for Kropotkin, as for Marx and Engels, was thus an organic or agency of class rule, the oppression of one class by another. Kropotkin was therefore always hostile to the notion that a socialist revolution could ever be achieved through the state (1993, 159–201).

Kropotkin rarely mentions Karl Marx in his writings, and when he does so, he tends to interpret him as a worshipper of the centralised state and as an advocate of the "volkstaat" (the folk or nation state). He even suggests that Marx is an exemplar of the Jacobin style of politics (Baldwin 1970, 50, 165). That Kropotkin understood Marx and his Social Democratic Party followers as state socialists is understandable when one considers Marx's own pronouncements on the essential role of the democratic state in the eradication of capitalism and the creation of a socialist/communist society.

In the famous *Communist Manifesto* (1848) Marx and Engels emphasised the aim of the Communist Party was the formation of the proletariat into a class and the "conquest of political power by the proletariat." This would involve, they suggest, the need to "centralize all instruments of production in the hands of the state ... centralization of credit in the hands of the state, by means of a national bank. .... Centralization of the means of communication and transport in the hands of the state. Extension of factories and instruments of production owned by the state. Establishment of industrial armies, especially for agriculture" (Marx and Engels 1968, 52–53).

Only when class divisions had been eradicated, and all production was in the hands of the nation (state) would public power lose its "political" character. This is the first intimation of the "withering away of the state" thesis.

Two years later (1850) in their address to the central committee of the Communist League, Marx and Engels again stressed that workers must strive not only for "the one and invisible German republic" but also "the most decisive centralization of power in hands of the state authority." They must also force the concentration of as many productive forces as possible—factories, transport, railways—in the hands of the state. It is thus the task of "the genuinely revolutionary party in Germany," they suggest, to "carry through the strictest centralization," and under no circumstances could the local autonomy of any village, town or province be tolerated (Marx 1973, 328–29).

The formation of the International Working Men's Association in 1864 was not, as Kropotkin stressed, the creation of Karl Marx, but rather the coming together of British trade unionists and the French followers of Proudhon—who, he suggests, were essentially mutualists (Baldwin 1970, 294). It was, as G.D.H. Cole put it, largely a trade union affair (1954, 88).

In a resolution as to the rules of the Association (1872) Marx and Engels reiterated that it was imperative that the proletariat form a political party to ensure the triumph of a social revolution, and this would entail the "conquest of political power" by the proletarian party (Kolpinsky 1972, 83).

The Paris Commune of 1871 also consisted of few Social Democrats (Marxists); the majority of members were followers of the French authoritarian socialist Auguste Blanqui, while the minority, largely members of the First International, were mutualists and followers of Proudhon's style of socialism—at least according to Engels (Marx and Engels 1968, 255). While Kropotkin always celebrated the Paris Commune he felt that it did not break with the tradition of the state, and of representative government, and thus was essentially a "failure." Kropotkin advocated a free commune, for he felt that the "emancipation of the workmen must be the act of the workmen themselves," as he put it in the 1887 article "Act for Yourselves" (1988, 32; Baldwin 1970, 155).

In contrast Marx and Engels heaped praise on the Paris Commune as a truly "national" and "working men's government"—a worker's state—and international in the sense of being the "glorious harbinger of a new society." It was, as Engels put it twenty years later, "the dictatorship of the proletariat" (Marx and Engels 1968, 259, 293–307). The phrase had earlier been used by Marx in his critique of the Gotha programme of the German Workers' Party (1875) in which he suggested that there would be a political transition period between capitalist and communist society, in which the state was "nothing but the revolutionary dictatorship of the proletariat" (327). Three years later Engels suggested that the revolutionary transformation would involve the proletariat seizing political power, turning the means of production into state property, the state itself representing the whole society. Then finally, Engels denotes, the government of persons would be replaced by the administration of things, and the state then simply "withers away" (Engels 1969, 333).

Gallons of ink have been expended on the question of what Marx and Engels meant by the "workers' state" or the "dictatorship of the proletariat."[6] Perhaps the orthodox Marxist Hal Draper, who devoted a whole book to the subject, expressed it succinctly when he wrote: "For Marx and Engels, from beginning to end of their careers and without exception 'dictatorship of the proletariat' meant nothing more and nothing less than 'rule of the proletariat'—the conquest of political power by the working class, the establishment of a workers state in the immediate post revolutionary period" (1987, 25).

Kropotkin, of course, argued that to hand over all the main sources of economic life to the state—whether or not this involved representative government or (as in Russia) a party dictatorship—would only lead to a new form of tyranny, that could only be described as "state capitalism" (Baldwin 1970, 286). Rudolf Rocker put it even more succinctly when he suggested that democratic socialism would inevitably lead to state capitalism (1978, 238).

5. Anarchist Political Strategies: Direct Action

Interpreting the state as a territorial institution that monopolises all forms of coercive power, and is invariably identified and sanctified by a national ideology, anarchists have always repudiated the state as an agency of social revolution. Kropotkin and other anarchists, both in their writings and practices, tended therefore to formulate essentially four alternative political strategies—alternatives, that is, to political action involving the state. Generally described as direct action, these are insurrectionism, anarcho-syndicalism, libertarian politics, and community activism.

In the early years of the anarchist movement, insurrectionism— or what became widely known as "propaganda by the deed"—was an intrinsic part of anarchism. But it was never more than a minority activity, although many well-known anarchists such as Errico Malatesta and Alexander Berkman were advocates of insurrectionism in their early years.[7] But both men, and the movement generally, quickly came to repudiate insurrectionism as a political strategy. The reasons for this were simple: it was elitist and alienated the majority of workers; it was ineffective, for rather than invoking a socialist revolution, it brought down the wrath of the state and the harsh repression of the whole anarchist movement. Although Kropotkin advocated a "spirit of revolt" he always repudiated individual acts of terrorism as a political strategy. A social revolution could be achieved, he felt, through a popular movement.[8]

In recent years insurrectionism has taken on a new lease of life, not in the form of assassinations, but in the form of protests and demonstrations. Seen as involving "direct action," anarchists have been conspicuous in the anti-globalisation demonstrations in Seattle (in 1999) and elsewhere. This has been interpreted as involving a resurgence of anarchism, and, moreover, as involving the emergence of a "new anarchism." According to Ruth Kinna this "new anarchism" consists of the following ideological categories: the anarcho-primitivism of John Zerzan; the acolytes of radical individualism of Max Stirner; the poetic terrorism (so-called) of Hakim Bey and John Moore, who follow the rantings of the reactionary philosopher Friedrich Nietzsche; the postmodern anarchism derived from Deleuze, Foucault, and Lyotard; and finally (believe it or not) the anarcho-capitalism of Murray Rothbard and Ayn Rand. What they have in common is that they all repudiated the "struggle by workers for economic emancipation" (Kinna 2005, 21).

As I have critiqued this so-called "new anarchism" elsewhere (2009), I will make only two points. Firstly, anarchists were in a distinct minority in these demonstrations, even if they were highlighted by the media who equated any violence that erupted with anarchism. For most of the participants were trade unionists, Marxists, and reformist liberals—like Naomi Klein—who seek to humanise capitalism and make it more benign. Secondly, the notion that anarchism had been in hibernation since the Spanish Civil War, only to emerge with the "new anarchism," is quite misleading. Anarchists (libertarian socialists) have been involved in many protests and demonstrations since the Second World War (and I speak from experience)—against the Vietnam War, against the apartheid system in South Africa, against specific environmental projects, against the pro-liferation of nuclear weapons, and the arms trade, and against the poll tax, as well as against global capitalism. Anarchism is alive (Gordon 2008)—it never died. And libertarian socialists, as fundamentally anti-capitalists, have been very much a part of the recent demonstrations.

Anarcho-syndicalism was the main strategy advocated by the anar-chists, and it is quite misleading to draw an absolute distinction between anarchist communism and anarcho-syndicalism. Although many revolu-tionary syndicalists were not anarchists, and tended to describe themselves as Marxists—Elizabeth Gurley Flynn, William "Big Bill" Haywood, and Daniel De Leon are examples—and although some anarchists repudiated anarcho-syndicalism, nevertheless anarcho-syndicalism was, and still is, an important anarchist strategy. Anarcho-syndicalism essentially emerged among the "Bakunist" or federalist section of the First International, and besides Kropotkin, many anarchists were fervent advocates of anarcho-syndicalism. They include, for example, Émile Pouget, Errico Malatesta, and Fernand Pelloutier. But the key figure in the development of anarcho-syndicalism was Rudolf Rocker, whose text *Anarcho-Syndicalism* (1938) has become an anarchist classic.

For Rocker anarcho-syndicalism was a form of libertarian socialism, its aim being the emancipation of the working class. It repudiated the tactic of "propaganda by the deed," opposed the parliamentary road to socialism of the German Social Democrats and the British Labour Party, which simply entailed reforms within the capitalist system, and, as with Bakunin and Kropotkin, it disavowed the Marxist conception of the "dic-tatorship of the proletariat" by means of a revolutionary political party, however this dictatorship was interpreted. For anarcho-syndicalists the trade unions and workers' associations had a dual function: to defend and improve workers' rights, wages, and living conditions in the present day—reforms from below; and to reconstruct social life through direct action, workers' solidarity and self-management, and federalist princi-ples. It considered that the workers' unions or syndicates, through engag-ing in class struggle, would become the "embryo" of a future socialist society. Anarcho-syndicalism thus repudiated "political action," by which they meant attempts to form a revolutionary government (as with the Marxists); it did not imply a denial of political or class struggle.[9]

But some anarchists have been critical of anarcho-syndicalism with its emphasis on the economy and workers' control, and as an alterna-tive strategy sought to develop explicit political institutions, as a counter-power to capitalism and the nation-state. They drew attention to, and inspiration from, the role of the *sans-culottes* and the *enragés* during the French Revolution, who through "sections" advocated a system of direct democracy; and also the importance of the soviets—self-managed popular assemblies—during the Russian Revolution. This political strategy envisaged an initial system of dual power, and the development within capitalism of popular assemblies, through which communities could directly manage their own affairs, via face to face democratic assemblies, or meetings, and co-federal organisations.

A harsh critic of anarcho-syndicalism, believing that it tended to uphold the factory system, Murray Bookchin was a strong advocate of this political strategy. He made an explicit distinction between democratic politics and statecraft, and came to formulate a form of communism, or political anarchism, that he described as libertarian municipalism (see Biehl 1998).

The final anarchist strategy is that of community activism, which may take many different forms. It is what the late Colin Ward described as "anarchy in action" (1973). It essentially involves ordinary people acting for themselves (as Kropotkin put it), taking control of their own lives, and through "direct action" establishing their own associations and groups independent of both the state and capitalism. It may involve squatting or establishing housing associations, the creation of food co-ops, affinity groups or independent schools,[10] or simply organising campaigns around environmental and community issues. The emphasis is on establishing voluntary associations that both enhance people's autonomy and reduce people's dependence on the state and capitalist corporations. Long ago Kropotkin emphasised the importance of such voluntary associations and mutual aid societies, some fleeting, some enduring, relating to many different domains of social life—artistic, political, intellectual, economic (Baldwin 1970, 132). He was fond of quoting the Lifeboat Association as an example of a spontaneous social organisation that was independent of the state and was motivated by mutual aid and not by the exploitation of others.

Indeed, Colin Ward defined anarchism as a social and political philosophy that emphasised the natural and spontaneous tendency of humans to associate together for their mutual benefit. Anarchism is thus the idea "that it is possible and desirable for society to organise itself without government" (1973, 12). Direct action within the community is thus something positive; it is not to be equated—as some do—with sabotage and with violent confrontations with the police over the occupancy of a building or a piece of land that has some iconic link to the capitalist system (which in any case, seems to be an ineffective strategy given the powers of the nation-state—as Malatesta learned long ago). Ward often uses the metaphor of the "seed beneath the snow" to suggest the kind of anarchist strategy that would enhance and develop all forms of mutual aid and voluntary cooperation within a community, small-scale initiatives that in some way undermined all forms of authority and the power of capitalism.

Such are the four political strategies, or forms of direct action, that have been articulated by revolutionary socialists, and Kropotkin is unique in putting an emphasis on all four strategies. All reflect the notion that a social revolution will not be achieved through a revolutionary vanguard party, or through the "parliamentary road" to socialism, but rather collectively through the activities of ordinary working people—by acting for themselves.

In an era of capitalist triumphalism, anarchism, libertarian social-ism, provides the only viable alternative to neoliberalism—for both liberal democracy (welfare capitalism) and Marxism (state capitalism) have been found wanting.

Notes

1   On the emergence of libertarian socialism or anarchism as a political movement and tradition see especially Baldwin (1970, 293–97), Pengam (1987), Cahm (1989), Nettlau (1996), and Van Der Walt and Schmidt (2009). Alain Pengam's notion that "Anarcho-communism has been regarded by other anarchists as a poor and despised relative" is quite misleading. It has been the main current.

2   For a clear and incisive argument that interprets anarchism as a historical movement and tradition that emerged in the late nineteenth century, and identifies it with anar-chist communism or libertarian socialism see Van Der Walt and Schmidt (2009).

3   For important and useful studies of Bakunin see Dolgoff (1973), McLaughlin (2002), and Leier (2006) as well as my own short introduction to his life and work (1993).

4   For studies of Kropotkin see Miller (1976, Purchase (1996), and Morris (2004).

5   For contemporary discussions of mutualism see Adán (1992) and Carson (2004).

6   Lenin's *The State and Revolution* (1949), written in 1917 shortly after his return to Russia, is largely an exposition of what he felt Marx and Engels meant by the "dictatorship of the proletariat."

7   For important studies of these anarchists see Richards (1965) and Fellner (1992).

8   On "propaganda by the deed" and insurrection see Morris (2004, 255–64), and Van Der Walt and Schmidt (2009, 128–33).

9   There have been many studies of revolutionary and anarcho-syndicalism: see, for example, Van Der Linden and Thorpe (1990). Noam Chomsky is the most well-known of contemporary anarcho-syndicalists. See the recent collection of essays and inter-views (Chomsky 2005). There are, of course, close affinities between anarcho-syndi-calists and council communism, which is essentially a libertarian form of Marxism (see Bricianer 1978, Shipway 1987, Pannekoek 2003).

10   Anarchists have always placed a great emphasis not only on political propaganda, but on education, and the schools of Francisco Ferrer in Spain, the Modern School Movement in the United States, and the free school movement are among the more well-known examples of anarchist initiatives (see Avrich 1980, Smith 1983).

References

Adán, José Pérez. 1992. *Reformist Anarchism*. Braunton: Merlin Books.
Avrich, Paul. 1980. *The Modern School Movement*. Princeton, NJ: Princeton University Press.

Baldwin, Roger N., ed. 1970. *Kropotkin's Revolutionary Pamphlets*. New York: Dover.

Biehl, Janet. 1998. *The Politics of Social Ecology: Libertarian Municipalism*. Montreal: Black Rose Books.

Bricianer, Serge. 1978. *Pannekoek and the Workers' Councils*. St. Louis: Telos Press.

Cahm, Caroline. 1989. *Kropotkin and the Rise of Revolutionary Anarchism, 1872–1886*. Cambridge: Cambridge University Press.

Carson, Kevin A. 2004. *Studies in Mutualist Political Economy*. Fayetteville, AR: n.p.

Chomsky, Noam. 2005. *Chomsky on Anarchism*. Oakland: AK Press.

Cole, G.D.H. 1954. *A History of Socialist Thought, Vol. 2: Marxism and Anarchism 1850–1890*. London: MacMillan.

Dolgoff, Sam, ed. 1973. *Bakunin on Anarchy*. London: Allen and Unwin.

Draper, Hal. 1987. *The "Dictatorship of the Proletariat" from Marx to Lenin*. New York: Monthly Review Press.

Engels, Friedrich. 1969. *Anti-Duhring*. London: Lawrence and Wishart.

Fellner, Gene, ed. 1992. *Life of an Anarchist: The Alexander Berkman Reader*. New York: Four Walls Eight Windows.

Gordon, Uri. 2008. *Anarchy Alive!* London: Pluto Press.

Kinna, Ruth. 2005. *Anarchism: A Beginner's Guide*. Oxford: Overworld.

Kolpinsky, N.Y., ed. 1972. *Marx, Engels, Lenin: Anarchism and Anarcho-Syndicalism*. Moscow: Progress Publishers.

Kropotkin, Peter. 1885. *Words of a Rebel*. 1992 edition. Montreal: Black Rose Press.

_____. 1892. *The Conquest of Bread*. 1972 edition. London: Penguin.

_____. 1899. *Memoirs of a Revolutionist*. New York: Grove.

_____. 1988. *Act for Yourselves*. London: Freedom Press.

_____. 1993. *Fugitive Writings*. Montreal: Black Rose Press.

Leier. Mark. 2006. *Bakunin: The Creative Passion*. New York: St. Martin's Press.

Lenin, Vladimir. 1949. *The State and Revolution*. Moscow: Progress Publishers.

Marshall, Peter. 2010. *Demanding the Impossible*. Oakland: PM Press.

Marx, Karl. 1973. *The Revolutions of 1848*. London: Penguin.

Marx, Karl, and Friedrich Engels. 1968. *Selected Works*. London: Lawrence and Wishart.

McLaughlin, Paul. 2002. *Michael Bakunin: The Philosophical Basis of His Anarchism*. New York: Algora.

Miller, Martin A. 1976. *Kropotkin*. Chicago: University of Chicago Press.

Morris, Brian. 1993. *Bakunin: The Philosophy of Freedom*. Montreal: Black Rose Press.

_____. 2004. *Kropotkin: The Politics of Community*. Amherst, NY: Humanity Books.

_____. 2009 "Reflections on the 'New Anarchism.'" *Social Anarchism* 42: 36–50; reproduced in this volume, 133–48.

Nettlau, Max. 1996. *A Short History of Anarchism*. London: Freedom Press.

Pannekoek, Anton. 2003 *Workers' Councils*. Oakland: AK Press.

Pengam, Alain. 1987. "Anarcho-Communism" In *Non-Market Socialism: The 19th and 20th Century*, edited by Maximilien Rubel and John Crump. London: MacMillan.

Purchase, Graham. 1996. *Evolution and Revolution: An Introduction to the Life and Thoughts of Peter Kropotkin*. Petersham NSN: Jura Books.

Richards, Vernon, ed. 1965. *Errico Malatesta: His Life and Ideas*. London: Freedom Press.

Rocker, Rudolf. 1938. *Anarcho-Syndicalism*. London: Secker and Warburg.

_____. 1978. *Nationalism and Culture*. St. Paul, MN: Coughlin.

Serge, Victor. 1963. *Memoirs of a Revolutionary*. Oxford: Oxford University Press.

Shipway, Mark. 1987. "Council Communism." In *Non-Market Socialism: The 19th and 20th Century*, edited by Maximilien Rubel and John Crump. London: MacMillan.

Van Der Linden, Marcel, and Wayne Thorpe, eds. 1990. *Revolutionary Syndicalism: An International Perspective*. Aldershot: Scolar Press.

Van Der Walt, Lucien, and Michael Schmidt. 2009. *Black Flame: The Revolutionary Class Politics of Anarchism and Syndicalism Vol. 1*. Oakland: AK Press.

Ward, Colin. 1973. *Anarchy in Action*. London: Allen and Unwin.

15

Anarchism, Individualism, and South Indian Foragers: Memories and Reflections (2013)

Prologue

This paper brings together two long-standing interests of mine, as reflected in two of my books, namely *Forest Traders* (1982), a study of the socioeconomic life of the Malaipantaram, a group of South Indian foragers, and *Kropotkin: The Politics of Community* (2004), which offers a critical account of the political philosophy and social ecology of the Russian revolutionary anarchist Peter Kropotkin. The books have one thing in common: both have been singularly ignored by academic scholars.

Apart from Peter Gardner (2000) no hunter-gatherers specialist—for example, Richard Lee, Tim Ingold, Alan Barnard, and Robert Kelly—ever mentions, or even cites, my study of the Malaipantaram. In a paper that emphasises the heterogeneous subsistence strategies among contemporary (not modern) hunter-gatherers, and aimed to mediate between the "traditionalists" (Richard Lee) and the "revisionists" (Edwin Wilmsen) in the rather acrimonious Kalahari debate, Nurit Bird-David argues that contemporary ethnographic inquiry should concern itself with the "dynamics of contact" between modern hunter-gatherers and capitalism at the *local level* (1992b, 21). This was precisely what my ethnographic study *Forest Traders*, published a decade earlier, had entailed! Anyone who reads this book will recognise that I don't treat the Malaipantaram as social isolates, nor do I deny them social agency—either individually or collectively—and I certainly do not describe the Malaipantaram as an unchanging society (Bird-David 1996, 260; Norstrom 2003, 49).[1]

Likewise, apart from a review in an obscure anarchist magazine, published by the Anarchist Federation, my book on Kropotkin has never been reviewed in any academic journal, or mentioned in recent academic texts on anarchism. A pioneer ecologist and a renowned geographer—his portrait still adorns the library of the Royal Geographical Society—Kropotkin was also an anthropologist, the author of the classic text on *Mutual Aid* (1902).[2] Yet the reviews editor of the *Journal of the Royal Anthropological Institute* did not consider a book on such an obscure Russian anarchist as having any interest to anthropologists. And given the current academic fashion with so-called post-anarchism, which invokes the ghosts of such radical individualists as Max Stirner and Friedrich Nietzsche, a book on Kropotkin seems of little relevance also to contemporary academic anarchists (e.g., Rouselle and Evren 2011).

The dean of Khoisan studies Alan Barnard once wrote a perceptive paper on primitive communism, aptly titled "Kropotkin Visits the Bushmen" (1993). In like fashion, in this paper I want to bring together my two books, thereby linking the Malaipantaram ethnography to Kropotkin's social anarchism, and to address the question: are South Indian foragers anarchists, or more precisely, can the social life of the Malaipantaram and other South Indian foragers be described as a form of anarchy?

Anarchy and Anarchism

I begin this discussion with a confession. When I completed a draft of my PhD thesis on the Malaipantaram in 1974—*Forest Traders* was a drastically shortened version of my thesis—I asked a close friend of mine, who had a degree in English literature and is a talented poet, to read the manuscript and to check it for any grammatical or stylistic errors or indiscretions. When she returned the thesis to me, she declared, "I don't believe a word of it." When I asked why, she responded: "It is just a description of your own anarchist politics. I think you've made the whole thing up."[3]

That hunter-gatherers, and tribal people more generally, have been described as living in a state of anarchy, has long been a common theme in anthropological writings. In recent years, with the highlighting of the presence of anarchists within the anti-capitalist and Occupy movement, together with the publication of James Scott's seminal *The Art of Not Being Governed* (2009)—subtitled *An Anarchist History of Upland Southeast Asia*—there has been an upsurge of interest in the relationship between anarchism and anthropology. Indeed. I long ago suggested that there was a kind of "elective affinity" between anarchism as a political tradition and anthropology. Scholars like Célestin Bouglé, Marcel Mauss, and A.R. Radcliffe-Brown had close associations with anarchism—Radcliffe-Brown in his early years was a devotee of Kropotkin and was known as "Anarchy Brown"—while anarchists such as Kropotkin, Élie and Elisée Reclus, Murray Bookchin, and John Zerzan have all drawn extensively on anthropological writings in developing their own brand of anarchism

(Morris 1998). Evans-Pritchard in his classic study *The Nuer* (1940) famously described their political system as "ordered anarchy," while Marshall Sahlins, equally famously, described the "domestic mode of production" as a "species of anarchy." In true Hobbesian fashion, Sahlins negatively portrayed the social organisation of tribal peoples as akin to a "sack of potatoes" (1972, 95–96; see Overing 1993 for an important critique of his neoliberalism). But more importantly, long before the current interest in anarchist anthropology, the Canadian scholar Harold Barclay wrote a perceptive little book, *People without Government* (1982), which is subtitled *The Anthropology of Anarchism*. The book affirms that "anarchy is possible" and describes not only hunter-gatherers—such as the Inuit, Bushmen, Yaka Pygmies, and the Australian Aborigines—as having anarchic societies, but also tribal societies more generally. Finally, we may note that Peter Gardner, in his pioneering study of the Paliyan foragers of South India, titles one of his key chapters "Respect, Equality, and Peaceful Anarchy," emphasising the fierce egalitarianism, the high value placed on individual autonomy, and the nonviolent ethos that pervades the social life of these foragers (2000, 83–100).

A distinction needs to be made, of course, between anarchy, which is an ordered society without government (or any enduring structures of domination and exploitation), and anarchism, which refers to a historical movement and political tradition. Emerging in Europe around the 1870s, in the aftermath of the defeated Paris Commune, mainly among workers in Spain, Italy, France, and Switzerland, anarchism as a political movement subsequently spread throughout the world in the early years of the twentieth century and is still a vibrant political tradition (Van Der Walt and Schmidt 2009).

In Kropotkin's own understanding of anarchist history, anarchism sought to actualise the rallying sentiments of the French Revolution— liberty, equality, and fraternity (social solidarity)—and entailed a creative synthesis of radical liberalism, with its emphasis on the liberty of the individual, and socialism (or communism), which implied a repudiation of capitalism, and the development of a society based on voluntary cooperation, mutual aid, and community life (Baldwin, 1970).[4]

Anarchism may be defined—at least for the purposes of the present essay—in terms of three essential tenets or principles.

Firstly, a strong emphasis is placed on the liberty of the individual. For the moving spirit of anarchism entails a fundamental focus on the sovereignty of the individual, and thus a complete rejection not only of the state power but all forms of hierarchy and oppression that inhibit the autonomy of the individual person. For social anarchists the individual was viewed of course as a social being, not as a disembodied ego or as some abstract individual, still less as some fixed, benign essence. A form of existential *individualism* is then a defining feature of anarchism as a political tradition.

Secondly, there is an emphasis on equality, and the affirmation of the community as a "society of equals." For anarchists this implied a complete repudiation of the capitalist market economy along with its wage system, private property, its competitive ethos, and the ideology of possessive individualism. Anarchism thus upheld *egalitarianism* as both a social premise, and as an ethical principle.

Thirdly, it expressed a vision of society based solely on mutual aid and voluntary cooperation, a community-based form of social organisation that would promote and enhance both the fullest expression of individual liberty, and all forms of social life that were independent of both the state and capitalism. The anarchists thus believed in voluntary cooperation, not in chaos, ephemerality, or "anything goes," and anarchists like Kropotkin viewed both tribal and kin-based societies as exhibiting many of the features of anarchy (Morris 2004, 173–90). *Communism*, or what Kropotkin described as free communism, was therefore one of the defining values (or characteristics) of anarchism as a political tradition—or at least a defining feature of the kind of anarchist communism that Kropotkin advocated (Morris 2004, 69–74).

Anarchists like Kropotkin therefore did not view anarchy as something that existed only in the distant past, in the Palaeolithic era, or as simply a utopian vision of some future society, but rather as a form of social life that had existed throughout human history—albeit often hidden in contemporary societies, buried and unrecognised beneath the weight of capitalism and the state. As Colin Ward graphically put it, anarchy is like "a seed beneath the snow" (1973, 11).

I turn now to the ethnography of South Indian foragers, and will address the question as to whether their social life can be described as a form of anarchy in terms of these three defining features of anarchism as a political tradition, namely, individualism, egalitarianism, and communism.

Individualism
It has long been recognised that the foragers of South India—such as the Kadar, Paliyan, Malaipantaram, and Jenu Kurubu—express what has been described as an individualistic ethic or culture (Gardner 1966, 408; Fox 1969, 145; Misra 1969, 234; Morris 1982, 109–10).

But to understand what this individualism entails, an initial note of clarification seems essential. For there has been a lamentable tendency on the part of many anthropologists to set up, in rather exotic fashion, a radical dichotomy between Western conceptions of the individual—misleadingly identified with Cartesian metaphysics or the "commodity" metaphor—and that of other cultures. This radical dichotomy suggests that Western culture views the human subject as an egocentric, isolated, nonsocial, and rigidly bounded individual, whereas in other cultures— specifically Bali, India, and Melanesia—people have a sociocentric concept of the subject, viewing people as intrinsically social beings, a "microcosm of relations," who conceive of themselves in terms of their social roles rather than as unique individuals (Geertz 1975, 360–411; Schneider and Bourne 1984, 158–99; Strathern 1988). Therefore, we are told, there exist "two types of person" (Carsten 2004, 84–88). This kind of dualistic approach is quite untenable, and has been subject to several telling criticisms (Spiro 1993; Morris 1994, 16–18; LiPuma 1998, 56–57). We need, in fact, to go back to Immanuel Kant, and to commonsense understandings of the world.

Kant famously described anthropology as the study of "what is a human being," or in contemporary parlance, "what it means to be human." In his seminal work *Anthropology from a Pragmatic Point of View* (2007), published in the last decade of his life, Kant suggested the understanding of the human subject in terms of a triadic ontology, viewing the individual as a universal human being (*mensch*), as a unique self (*selbst*), and as a social being, a member of a particular group of people (*volk*).[5] Kant, of course, focused on the individual as a universal, rational subject, while his student and later critic Johann Herder stressed that humans were fundamentally cultural beings. Many years later, the anthropologist Clyde Kluckhohn expressed this triadic ontology in simple everyday terms. Critical of dualistic conceptions of the human subject in terms of the nature (biology) versus culture dichotomy, Kluckhohn suggested, in contrast, that there were three essential ways of understanding the human subject. Every person, he wrote, is in some respects like every other human being—as a species being (humanity), that they are like no other human being in having a unique personality (or self); and, finally, that they have affinities with some other humans, in being social or cultural beings, or persons (Murray and Kluckhohn 1953).

This relates, of course, to the fact that all humans are embedded in three interrelated historical processes, involving distinct geo-temporal levels or realms, namely, the phylogenetic, relating to the evolution of humans as a species-being (our *human* identity); the ontogenetic, which relates to a human person's own life-history as a unique individual or embodied self, situated within a specific and ecological setting (our *self* identity); and the socio-historical, which situates the human subject as a person within a specific social and cultural context (expressed in various *social* identities, which in all cultures are multiple, shifting and relational). The "holistic person" (Riches 2000) is, of course, acknowledged in most human societies.

The notion that humanity is simply a class concept or does not exist—only individual humans, we're told, exist—as expressed by John Gray (2002, 12)[6] and some anthropologists, is quite misleading. As Manuel DeLanda argues, humanity (as well as other species-beings) do indeed exist, but at a different geo-temporal scale to that of the organism—the unique human individual. For a species-being like humanity comes into being, endures as an entity over perhaps thousands or even millions of years, eventually to be transformed, or become extinct, as with the dodo and great auk (DeLanda 2006, 48–49).

The Malaipantaram, and other foragers of South India, like people everywhere, clearly affirm this triadic ontology. They recognise that the people they encounter are human beings (manushyan), as distinct from elephants and monitor lizards, and that they have unique personalities, and a sense of their own individuality (which ought not to be equated with individualism as a cultural ethos). They recognise too that other humans, both within their own society and with regard to outsiders, are social beings with ethnic affiliations and diverse social identities (Rupp 2011, 14–17).

But what is significant about the Malaipantaram and other South Indian foragers' conception of the human subject is that they place—like the existentialists, but unlike evolutionary psychologists and Durkheimian sociologists—a fundamental focus on the individual as an independent person. A fundamental value and stress is put on the autonomy of the individual. Both in terms of their child-rearing practices and their gathering economy, the Malaipantaram place a high degree of emphasis on the person as an existential being—as self-reliant and autonomous. Socialisation patterns—as a dialectical process—are largely geared to making a child socially, psychologically, and economically independent at a very early age, and to respect the autonomy of other individuals. Thus in contrast with neighbouring caste communities the Malaipantaram individual may constitute a unit of both production and consumption; and to live a solitary existence is not only possible but by no means unusual for older men, and it is not portrayed negatively (Morris 1982, 140–47). A strong adherence to individual autonomy, and thus an "intense" individualism (as it has been described) has been recognised by all researchers on South Indian foragers.

Although clearly linked to a gathering economy, both for subsistence and trade, such individualism is also expressed in the diversity of their economic strategies, and their individual mobility and flexibility with regard to group membership. Indeed, individualism has been interpreted as a characteristic feature of hunter-gatherer societies generally, or at least those with an immediate-return economic system (Woodburn 1982).

Nobody has expressed this individualistic ethos with more cogency than Peter Gardner, who in relation to the Paliyan and several other foragers views it as intrinsically linked to individual decision-making, social mechanisms that undermine any form of hierarchy, a nonviolent ethos (in that interpersonal conflicts are generally resolved through fission and mobility), and a general absence in the formalisation of culture (Gardner 1966; 1991, 547–49; 2000, 83–85).

In *Forest Traders* I noted that the ethnographic data on the Paliyan, who were characterised by a "very extreme individualism" (Gardner 1966, 409), seemed to run counter to Louis Dumont's argument in *Homo Hierarchicus* (1970), wherein he suggested that the presence of the "individual"—in its modern, normative sense—was not recognised in Indian society. I have written elsewhere on this apparent paradox (Morris 1978; 1991, 262–69), but what has to be recognised in the present context is that not only must a distinction be made between individuality (and the agency of the individual—acknowledged in most societies) and normative or cultural conceptions of the human subject, but also the fact that there are many distinct and contrasting forms of *individualism* (Lukes 1973). Dumont himself devotes a good deal of discussion to two forms of individualism, besides that of economic individualism. One is that of the ascetic "renouncer" in Hindu society, the *sannyasin*, whose individual identity is achieved by repudiating all ties that bind a person to the caste system (society) and the world. The other form of individualism is that associated with the concept of *bildung* or "self-cultivation" that was particularly associated with literary intellectuals in Germany at the end of the eighteenth century, and was later developed by Nietzsche and by his poststructuralist devotees (Dumont 1986a, 1986b).[7]

But the key distinction that has to be made is that between the individualism of the South Indian foragers and the various kinds of individualism that are generally associated with the capitalist economy, if not with many aspects of Western culture. These range from that of Cartesian philosophy, with its notion of a disembodied ego radically separate from nature and social life (critiqued by Levi-Strauss, 1977, 41);[8] the abstract or possessive individualism of liberal theory that was long ago lampooned by Marx and Bakunin; the methodological individualism of optimal foraging theory (critiqued by Ingold, 1996); and the radical egoism of Ayn Rand which advocates a form of selfishness that has little or no regard for other humans and is now apparently dominant in right-wing American politics (Weiss 2012, 9–10).

All four kinds of bourgeois individualism were repudiated by Kropotkin and the early social anarchists, as all tended to deny the social nature of the human subject. The mode of individualism that Kropotkin advocated, in contrast, he described as *"personalismus or pro sibi commu-nisticum"*—the kind of individuality that is inherently social (Kropotkin 1970, 297).

In a more recent text Susan Brown has followed Kropotkin in making a clear distinction between instrumental or possessive individualism, manifested through the market, and the existentialist individualism advocated by the social anarchists—as expressed in the individual's capacity to be autonomous and self-determining (Brown 1993, 32–33; Morris 2004, 183–86). Clearly, the kind of individualism expressed by the Malaipantaram and other South Indian foragers has close affinities with the existentialist individualism described by Brown. Indeed she defines social anarchism as combining existential individualism with free communism (1993, 118).

But, of course, making a distinction between the individualism of hunter-gatherers and the rugged individualism of modern capitalism is not saying anything new or original. For long ago Stanley Diamond emphasised that the mode of thought expressed by tribal societies, specifically hunter-gatherers, with respect to the human subject, was one that was concrete, existential, and personalistic (Service 1966, 83; Diamond 1974, 146).

Egalitarianism

In *Forest Traders* I emphasised that an egalitarian ideal permeated Malaipantaram society, and that this ethos contrasted markedly from the emphasis on hierarchy in surrounding agricultural communities, particularly in relation to gender and with regard to the higher castes (1982, 49). In fact, I argued that Malaipantaram society was both egalitarian and individualistic in its essential ethos (110). Around the same time James Woodburn published his seminal paper on "Egalitarian Societies" (1982), although it is well to recall that as a concept "egalitarianism" is hardly mentioned in what constitutes one of the founding texts of hunter-gatherer studies (Lee and DeVore 1968). But it is generally recognised that egalitarianism is a fundamental characteristic not only of South Indian foragers but of all hunter-gathering societies with an immediate-return economic system (Gardner 1991, Boehm 1999, Widcok and Tadesse 2005, Solway 2006).

Among the Malaipantaram and South Indian foragers egalitarianism is manifested in diverse ways, as an ethos, as a constellation of values and normative expectations, and in their social relationships. Indeed, their cultural values and social actions are interlinked and dialectically related. We can briefly mention, with regard to such egalitarianism, three topics of interest: the emphasis on sharing, their attitude towards authority structures, and their general emphasis on equality, especially in relation to gender.

Like many other hunter-gathering societies, the Malaipantaram always share meat from wild animals within the camp or settlement and every person is entitled to a share. Although the hunting of mammals such as sambar, chevrotain, and various monkeys is practised, most of the meat obtained is through eclectic foraging, and what almost amounts to the "gathering" of small animals—specifically tortoises, bats, monitor lizards, and squirrels (Morris 1982, 71–79). Apart from the sharing of meat, sharing is in fact limited among the Malaipantaram, and economic exchanges within the community is best described in terms of reciprocity and mutual aid.

Although in various Malaipantaram settlements there are recognised "headmen" (*Muppan*), these are largely a function of administrative control introduced by the state via the Forest Department, in order to facilitate communication. Such headmen have little or no control over the lives or movements of other members of the local group (Morris, 158–59). There is, in fact, among the Malaipantaram and other South Indian foragers a marked antipathy towards any form of authority or hierarchy, whether based on wealth, prestige, or power. In their everyday life the emphasis is always on being modest, non-aggressive, and non-competitive, engaging with others in terms of an ethos that puts a fundamental emphasis on mutual aid and on respecting the autonomy of others, especially those with whom a person regularly associates. But as many scholars have indicated, the stress on egalitarian relationships does not simply imply a lack of hierarchy, but is actively engendered by forms of social power—diffuse sanctions or levelling mechanisms—expressed and enforced by means of criticisms, ridicule, ostracism, desertion, or by the simple, voluntary adherence to customary norms—as I am doing in writing this essay in intelligible English. (On diffuse sanctions among hunter-gatherers see Lee 1979, 457; Barclay 1982, 22.) It seems to me that to describe this form of social power, and the diffuse sanctions that are entailed, as implying "reverse-dominance hierarchy" (Boehm 1999, 86–88) or as a form of "governance" (Norstrom 2001, 207) is quite misleading, for the Malaipantaram have a marked aversion to all forms of domination and governance; this, of course, does not imply that they live in the forest in a state of "anomie." They are, however, like other South Indian foragers, well attuned to the "art of not being governed" (Scott 2009).

Given this emphasis on egalitarian relations it is not surprising that gender relations among the Malaipantaram and other South Indian foragers, are typically recognised as being one of equality. In *Forest Traders*, acknowledging the social and economic independence of women, I stressed that there was a high degree of equality between the sexes, especially when contrasted with gender relations among caste communities.

Within the conjugal family men and women have more or less equal rights, and neither party has authority over the other (1982, 40). But given the harassment that Malaipantaram have generally to contend with, or what Gardner described as "inter-cultural pressure" (1966, 391), I never encountered a Malaipantaram woman foraging alone in the forest (on gender relations among South Indian foragers see Ehrenfels 1952, 209; Mathur 1977, 178; Bird-David 1987, 154; Gardner 2000, 101–3).

In a widely acclaimed essay on Marshall Sahlins's renowned thesis describing hunter-gatherers as the "original affluent society" (1972, 1–39), Nurit Bird-David (1992a) seems to inflate the notion of sharing, which is a key element of egalitarianism, into a rather metaphysical principle. Sharing thus becomes a totalising concept that incorporates almost all aspects of the social and cultural life of the Nayaka. By implication her "model" applies to all foragers with an immediate-return economic system, although interestingly she makes no mention at all of other South Indian foragers. It is of interest, too, that whereas Bird-David *endorses* Sahlins's thesis of the hunter-gatherer "Zen road to affluence," in *Forest Traders* I made no mention at all of this questionable thesis (see Kaplan 2000 for a measured critique of its limitations).

There is, of course, nothing amiss with describing South Indian foragers as individualistic or egalitarian, or as having a "sharing ethos," but Bird-David's analysis is a typical example of what Kathleen Morrison describes as typological and essentialist thinking (2003, 3) for although this analysis obviously has a ring of truth, it gives quite a biased portrait of the social and cultural life of South Indian foragers. For in following Colin Turnbull's (mis)interpretation of Mbuti religion as a form of crude pantheism, Bird-David tends to *conflate* natural phenomena—whether mountains, rivers, rock outcrops, stones, trees, or animals (especially elephants)—as well as artefacts with the spiritual beings—malevolent spirits, ancestral spirits, and forest deities—that, according to the forager's religious ideology, inhabit or have their "abode" in the forest, or are identified with certain figurines or icons. But these two aspects of the forager's life-world are distinct, and Ananda Bhanu writes, for example, of the Chola Naickan calling on the spirits, represented in an elephant figurine (*aneuruva*) to protect them, not only against illness and misfortunes, but from the maraudings of real elephants (1989, 63–66). Indeed, rather than living in a "giving environment" (Bird-David 1990) the Malaipantaram and other South Indian foragers appeal to the ancestral spirits (*chavu*) or the forest deities (*malai devi*, mountain gods) not only when there are illnesses and misfortunes, but also when there is a lack of food or honey or hunting has not been successful. Through shamanistic rituals the reasons for the lack of food or misfortunes is explained in terms of the foragers not upholding certain customary norms or moral edicts, particularly not respecting other people's autonomy, or not sharing or offering mutual aid to their close kindred. Equally, by focusing on their religious ideology Bird-David completely ignores the empirical naturalism that is manifested in their ecological knowledge of the natural world and in their practical activities, and which is equally a part of the foragers' culture. The metaphor "giving environment" is misleading,[9] for like people in all societies the relationship of the Malaipantaram and other South Indian foragers to the natural world is one that is complex, diverse, multifaceted and often contradictory, and cannot be reduced to a single metaphor, however engaging and illuminating.

The Malaipantaram, of course, recognise the forest as their home, and have warm feelings towards the forest as the essential source of their livelihood and well-being. It is the abode of their ancestral spirits and the forest deities on whom they can always call upon for support and protection, and thus the forest is always a place of refuge (Morris 1982, 50). But it is misleading to interpret the forest simply as a "parent" or "giving" for the foragers.

Likewise, although sharing is a fundamental ethic among the Malaipantaram one has to recognise that it is not ubiquitous. For the Malaipantaram are involved in market relations, and though they always strive to make these relationships inclusive—as Justin Kenrick (2005, 135) describes it—that is friendly, sharing, personal and one involving mutual aid and support—these relationships are essentially hierarchical and exploitative, and they are ones that the foragers do not control. As Mathur describes market relationships among the Chola Naickan—who are clearly related to the local group described by Bird-David—what the foragers attempt and prefer is to "act as a generous kinsman rather than a self-seeking trader"—in contrast to the forest contractor who always seeks to control and exploit the foragers (1977, 36).

Equally important, one has to recognise that there is very little sharing between families among the Malaipantaram, and it is the family that is the key productive and commensal unit. It always struck me as unusual that after a day spent together in eclectic foraging, both for food and for minor forest produce, cooperating, conversing, and engaging in banter, on returning to the cave in the late afternoon each of the three families would build their own fireplace, establish a distinct commensal unit, and that there would be no sharing at all between the different families.

Although, of course, a distinction can be made between tribal agriculturalists and South Indian foragers like the Malaipantaram—and even between those Malaipantaram who are settled cultivators and those foragers living exclusively in the interior (on whom I focused in my research), it is quite misleading to set up a radical opposition between sharing and reciprocity. All human societies engage to some degree in sharing or generalised reciprocity (Price 1975), but the relationship between such sharing and reciprocity is always dialectical. Malaipantaram men who are deeply involved in hunting, especially in the marketing of the Nilgiri langur, tend to live separately with their family, in order to avoid sharing the meat, while any individual who constantly engages in demand sharing—without any reciprocation—is likely to find his or her own kin moving elsewhere. And among the Malaipantaram and other South Indian foragers there is a clear distinction made between food gathered from the forest, and the goods—whether rice, condiments, or artefacts—that are obtained from engaging in market relations. The latter kind tend to involve a more balanced reciprocity (see Norstrom 2003, 221). The key idea expressed by the Malaipantaram is one of mutual aid, which includes sharing, reciprocity, and an ethic of generosity (Morris 1982, 161).

Communism
The American poet Kenneth Rexroth once wrote: "People who hunt and gather cannot be anything but communist" (1975, 1).

There has, of course, been a long tradition, going back to Lewis Morgan and Friedrich Engels in the nineteenth century, affirming that hunter-gatherers—those "roving savages" as Engels described them—and tribal or kin-based societies more generally, live in "communistic communities" (Marx and Engels 1968, 579). By communism was meant not simply the absence of private (exclusive) property but rather a universal collective right of access to all resources necessary for life and well-being, specifically rights to land and the means of production; what property was "owned" being of a purely personal nature (Leacock 1983).

In his well-known reflections on "primitive communism" Richard Lee (1988) noted that the concept was not only ignored by later Marxists—Chris Harman's A People's History of the World (1999), for example, barely recognises the existence of tribal peoples—but was generally dismissed and belittled by most anthropologists. Lee tends to identity primitive communism with what he describes as the "foraging" mode of production. He tends, therefore, to leave out of his account not only hunter-gatherers that are sedentary and non-egalitarian—such as the Ainu (Japan), Calusa (Florida), and the hunter-gatherers of the northwest coast of the Americas—but also the Iroquois. For Morgan, however, it was the Iroquois that were the prototypical primitive communists. Democratic and egalitarian, they were the exemplars of the "liberty, equality, and fraternity of the ancient gentes" (Morgan 1877, 562).

In several essays Lee outlines what he terms the "core features" of primitive communism, or the foraging mode of production. These are the following:

- Social life in the society or ethnic community is focused around a band structure, small groups of twenty to thirty people, with highly flexible membership and an emphasis on mobility, often with patterns of concentration and dispersal.
- Land is held collectively in common and everyone has free or reciprocal access to resources. There is no ownership in the sense of completely withholding access.
- There is an ethic of egalitarianism, and this implies strong, even if diffuse, sanctions against the accumulation of wealth, all forms of political authority, and any expression of self-aggrandisement.
- Such egalitarianism implies an emphasis on cooperation and mutual aid, and patterns of sharing or generalised reciprocity are strongly emphasised, especially within the band or camp. No one goes hungry if food is available.
- An emphasis on cooperation and sociality is combined with a high respect for the autonomy of the individual. There is therefore a strong emphasis on individual choice with regard to whom a person resides or associates with (Leacock and Lee 1982, 7–8; Lee 1988, 254; 2005, 20).

Primitive communism or the foraging mode of production thus entails a combination of individualism and egalitarianism. What is evident, of course, is that these core features or attributes of the foraging mode of production—primitive communism—are virtually synonymous with those specified by Woodburn (1982) in terms of immediate-return economic systems, by Gardner (1991) in terms of the individual autonomy syndrome, and Barnard's (2001) specifications of the hunter-gatherer "mode of thought."

What is equally significant is that the Malaipantaram, as depicted in my study *Forest Traders*, as well as South Indian foragers more generally, can clearly be described as communists, as exemplars of the foraging mode of production (or thought), even though they may engage in a diversity of economic activities (Fortier 2009, 102).

It has become rather fashionable nowadays to suggest like Ayn Rand—Margaret Thatcher's guru—that societies do not exist, and that all we supposedly experience is sociality, social networks or lines, with the human person simply being the nodal point or intersection of various relations; though nobody seems to doubt the reality of the state and capitalist organisations. It has even been suggested that the concept of society is not applicable to hunter-gatherers, the concept being defined, in rather

Hobbesian fashion, as entailing "structures of domination." The concept of society, of course, emerged at the end of the eighteenth century, as a relational concept, in contradistinction to the nation-state, that was then asserting its hegemony. The distinction between society and the state (entailing structures of political domination or government) was not only articulated by such radical liberals as Tom Paine but, of course, by numerous anarchists from William Godwin at the end of the eighteenth century to Kropotkin, Reclus, and the social anarchists in the twentieth century.

Like other primates, all humans live in groups of various kinds, and from the ethnographic record it is quite clear that hunter-gatherers live in societies. For the Hadza have a clear and distinctive ethnic identity (Woodburn 1988, 39); the Jo'hoansi of the Kalahari recognise themselves as a society of "real people" with a distinctive culture and language, as many scholars have described (Kent 2002); the Mbendjele of central Africa regard themselves both as Yaka (or forest people) as well as a distinct society (Lewis 2005, 57); and, finally, with regard to the Paliyan of South India, they are described as being "both in their own, and in their neighbours' eyes, a separate ethnic group" (Norstrom 2003, 4) or in everyday language, a society. Dismissing the concept of "society" as a "stereotype" is hardly helpful to understanding hunter-gatherers.[10]

A human society, of course, is simply a group of humans who are linked to each other by enduring social ties, who share certain values, beliefs, normative expectations, and purposes, and whose members have a sense of belonging to the group. A society, as Kropotkin long ago emphasised, is essentially a moral, not a political, community (Boehm 1999, 12). To even talk about an "egalitarian ethos" or "mode of thought" presupposes the existence and identity of a specific hunter-gatherer society.

The Malaipantaram, and other South Indian foragers, identify themselves as belonging to a specific society or ethnic community, and are recognised as such not only by the Indian state, but by all people with whom they come into contact. But they do not form a cohesive political unit, but are "fragmented"—as I expressed it in *Forest Traders* (1982, 112)—into diverse, flexible but interrelated social groupings. Basically, with regard to the Malaipantaram, as a society or moral community, there exist three levels of organisation, namely the forest settlements, the conjugal family, and what I termed forest camps, the loose groupings of two or three families, who are linked essentially by affinal ties (1982, 30).

Named after particular forest locations, the dispersed *settlements* are local groups, consisting of between eighteen and thirty-six people. They are situated in specific valleys or forest ranges. At the settlements the Malaipantaram often engage in the cultivation of tapioca or rice, as well as certain fruit-bearing trees, and often engage in casual agricultural labour on nearby estates. People at the settlement are intimately associated with the hill forests above the settlement, and have extensive knowledge of the forest environment and its diverse inhabitants. They also form ritual congregations in relation to the ancestral spirits (*chava*) and mountain deities (*malai devi*) which have their "abodes" in the same forested hills. The settlement or local group is rarely a cohesive unit, and is often divided into several "clusters" or hamlets, each composed of two to four families (see Misra 1969, 206; Bird-David 1983 on the Jenu Kuruba and Nayaka, respectively). Though widely dispersed over the Ghat mountains south of Lake Periyar, the settlements are by no means isolated, for all Malaipantaram are linked by a universal system of kinship (Barnard 2001, 94), and thus there are kinship links, particularly affinal ties, between members of the various settlements.

In contrast the conjugal family is a fairly cohesive unit, forming a distinct social and economic grouping. As noted above, the family is the basic productive and commensal unit, even though marriage ties are often transient. Conjugal relations are generally warm and affectionate, essentially reciprocal, and complementary.

Between the dispersed settlements or local groups, and the conjugal family there are social aggregates that are difficult to define. But the fact that the Malaipantaram do not have corporate, land-owning kin groups, is no reason to suggest that they do not live in groups! I have discussed in *Forest Traders* at some length the forest groups of the Malaipantaram and have suggested that although membership is often flexible and transient, all Malaipantaram are highly mobile, acting as independent persons, such groups are in the nature of a kindred, and people are united in terms of affinal ties. As I put it: "Affinal links seem to serve as a guiding principle in structuring friendships and camp aggregates" (1982, 157).

People that constitute the camp aggregates are drawn together by "dyadic bonds of affection" and express enduring relationships of mutual aid, as well as expressing an ethic of sharing and reciprocity. Malaipantaram social structure cannot therefore be interpreted simply as consisting of independent families with floating "single persons" (mostly adolescents) giving social cohesion to the local group. All Malaipantaram act, and are expected to act, as autonomous individuals; it is affinal ties that structure their relationships and their forest camps (see Bird-David 1987, Demmer 1997).

Malaipantaram social life, and that of other South Indian foragers, can therefore be described, in important respects, as communistic. But many scholars have suggested that there is an inevitable tension, or contradiction, between egalitarianism (with the emphasis on mutual aid, sharing, and sociality) and individualism (with stress on the autonomy of the individual) (Riches 2000, 671–72). Indeed liberal scholars like Isaiah Berlin stressed that there was an inherent conflict between the values of equality and liberty. But as anarchists have always emphasised, the two concepts are dialectically related, and necessarily imply each other. As Kropotkin put it, you can hardly be free and independent in a society based on inequality and hierarchy.

Among the Malaipantaram, and it seems hunter-gatherers more generally, an egalitarian ethos seems to coexist with an equal emphasis on the autonomy of the individual. As I wrote long ago: "A viable ethnographic portrait can be drawn only if we stress the co-operation and the individualism, the warm attachments Malaipantaram hold towards each other and the fragility and ephemeral nature of these ties" (1982, 114), which reflects in particular the emphasis placed on the autonomy and self-sufficiency of the individual.

Conclusion

Having outlined above the individualism, egalitarianism, and communism that is undoubtedly manifested in the social life of the Malaipantaram and other South Indian foragers, can such foragers be described as living in a state of anarchy?

Judging by the ethnographies of Peter Gardner (2000) and myself (1982), the answer seems to be in the affirmative. Indeed Jana Fortier certainly suggests that the micropolitics of both the Paliyan and the Malaipantaram can be depicted as anarchy (2009, 103). Yet other scholars have baulked at the idea. Christopher Boehm, for example, dismisses the whole idea that the "egalitarian blueprint" of hunter-gathering is a form of anarchy, suggesting that anarchy implies a complete absence of power and control (1999, 87). Apart from Stirnerite individualists, no social anarchist has ever envisaged a society without some form of social power, immanent within the community. Power as mutuality and equality, as Barclay expressed it (2009, 7–26).

In my study of the Malaipantaram, written almost forty years ago, I emphasised how much of their social life and culture—their nomadism, their flexible organisation, and their reluctance to take up agriculture and become a settled community—was an attempt to retain their autonomy as a forest people. Thus the Malaipantaram attempted to retain their independence and cultural integrity despite being ridiculed and harassed by the caste communities of the plains, the exactions of the Indian state bent on their development and settlement, and the intrusions of a mercantile economy focused around the trade of minor forest products. The suggestion that in emphasising the external social factors that impinge on Malaipantaram social life, I thereby deny them social agency (Nostrom

2003, 48–51) is quite misleading. The whole ethnography is, in fact, about their social agency: how the Malaipantaram derive their basic livelihood, how they organise their kin relations and forest camps, and how through their nomadism and flexible social organisation—and many other ploys— they retain a sense of autonomy and independence. And I emphasised too, like many other late scholars of hunter-gatherers, that among the Malaipantaram a strong emphasis on personal autonomy and independence coexisted, as I noted above, with an equal emphasis on mutual aid, sharing, and egalitarian relationships. Malaipantaram social life could therefore be described as a form of anarchy.

Anarchists like Kropotkin and Reclus, of course, always recognised and affirmed that the basic principles of anarchism—the liberty of the individual, egalitarianism, and a form of social life based on cooperation, sharing, and mutual aid—were characteristic of hunter-gatherers and tribal society generally (Kropotkin 1902, 74–101; Morris 2004, 173–90). But what they advocated was not a return to hunter-gathering and some form of anarcho-primitivism (Zerzan 1994) but rather, drawing on the knowledge, technics, arts, and sciences that humanity has accumulated over the past five thousand years or so, their aim was to engender an anarchist-communist society, where productive activities and all social functions would be organised through voluntary cooperation and mutual aid, the wealth produced being shared equally with *all*.

Notes

1   In fact, a decade before the revisionist controversy, I was emphasising, while explicitly acknowledging the earlier seminal work of Richard Fox (1969) and Peter Gardner (1972), that the Malaipantaram had long been incorporated into a wider mercantile capitalist economy. But in dialectical fashion, I also stressed their social agency in maintaining their autonomy and cultural integrity as a foraging community.

2   Essentially this text was a repudiation of the Social Darwinism of Herbert Spencer and Thomas Huxley that was prevalent at the end of the nineteenth century.

3   This, of course, raises the interesting question as to what degree anthropologists, in their ethnographic accounts, impose upon the data their own epistemological and political preconceptions?

4   This synthesis is well illustrated by Michael Bakunin's familiar adage: "That liberty without socialism is privilege and injustice, and that socialism without liberty is slavery and brutality" (Bakunin 1973, 110).

5   By "pragmatic" what Kant intended was the use of such knowledge to further human enlightenment, to widen the scope of human freedom, especially from religious dogma and political oppression, and thus to advance the "dignity" of humans.

6   This does not stop Gray from defining humanity as *Homo rapiens*—the destructive primate (2002, 151).

7   Recognising the many forms of individualism does not entail, of course, that we must follow the facile practice of postmodern anthropologists and put the concept in inverted commas (quotation marks).

8   The tendency of many postmodern anthropologists to equate Western culture with Cartesian metaphysics and mechanistic philosophy is, of course, reductive and facile.

9   Describing the natural environment (forest) as "giving" is akin to describing Margaret Thatcher as an iron-lady. Although such a metaphor, or "poetic evocation," has a certain truth and validity, it is quite misleading in that it completely obscures the complexity of Thatcher's politics. And nobody, of course, thought Thatcher was actually made of iron. Likewise, the metaphor "giving environment" or "forest is parent" is restrictive and limiting—and obscures the complexity of the foragers' relationship with the forest environment.

10   It is often said that foragers (or tribal people) do not have a concept of "society" that matches that of the anthropologists. Why on earth should they? Foragers may not have a concept of the "economy" or "culture" but this does not imply that they have no economic life or no culture. Terms such as *kudumbam* (family, kin group) like the English terms "group" or "society" have a wide range of meanings (or referents). This does not imply they have no society, or that the Malaipantaram are not a society.

References

Bakunin, Michael. 1973. *Selected Writings*. Edited by Arthur Lehning. London: Cape.

Baldwin, Roger N. 1970. *Kropotkin's Revolutionary Pamphlets*. New York: Dover.

Barclay, Harold. 1982. *People without Government: The Anthropology of Anarchism*. London: Kahn & Averill.

_____. 2009. *Fresh Alternatives*. Vernon, British Columbia: Lake City Printers.

Barnard, Alan. 1993. "Primitive Communism and Mutual Aid: Kropotkin Visits the Bushmen." In *Socialism: Ideals, Ideologies and Local Practices*, edited by C.M. Hann, 27–42. London: Routledge.

_____. 2001. *The Hunter-Gatherer Peoples*. Buenos Aires: Fundacion Navarro Vida.

Bhanu, B. Ananda. 1989. *The Cholanaickan of Kerala*. Calcutta: Anthropological Survey of India. Memoir no. 72.

Bird-David, Nurit. 1983. "Wage Gathering: Socio-Economic Changes and the Case of the Food-Gathering Naikens of South India." In *Rural South Asia: Linkages, Change, and Development*, edited by Peter Robb, 57–88. London: Curzon Press.

_____. 1987. "Single Persons and Social Cohesion in a Hunter-Gathering Society." In *Dimensions of Social Life*, edited by Paul Hockings, 151–65. Berlin: M. de Gruyter.

_____. 1990. "The Giving Environment: Another Perspective on the Economic System of Gatherer-Hunters." *Current Anthropology* 31, no. 2: 189–96.

_____. 1992a. "Beyond 'the Original Affluent Society': A Culturalist Reformulation." *Current Anthropology* 31, no. 1: 25–47.

_____. 1992b. "Beyond the Hunting and Gathering Mode of Subsistence: Observations on the Nayaka and Other Modern Hunter-Gatherers." *Man* 27, no. 1: 19–44.

_____. 1996. "Puja, or Sharing with the God? On Ritualized Possession among the Nayaka of South India." *Eastern Anthropology* 49: 259–76.

Boehm, Christopher. 1999. *Hierarchy in the Forest: The Evolution of Egalitarian Behavior*. Cambridge, MA: Harvard University Press.

Brown, L. Susan. 1993. *The Politics of Individualism: Liberalism, Liberal Feminism, and Anarchism*. Montreal: Black Rose Books.

Carsten, Janet. 2004. *After Kinship*. Cambridge: Cambridge University Press.

DeLanda, Manuel. 2006. *A New Philosophy of Society: Assemblage Theory and Social Complexity*. London: Continuum.

Demmer, Ulrich. 1997. "Voices in the Forest: The Field of Gathering among Jenu Kurumba." In *Blue Mountains Revisited: Cultural Studies on the Nigiri Hills*, edited by Paul Hockings, 164–91. Delhi: Oxford University Press.

Diamond, Stanley. 1974. *In Search of the Primitive: A Critique of Civilization*. New Brunswick: Transaction Books.

Dumont, Louis. 1970. *Homo Hierarchicus: The Caste System and Its Implications*. London: Weidenfeld & Nicolson.

_____. 1986a. *Essays in Individualism: Modern Ideology in Anthropological Perspective*. Chicago: University of Chicago Press.

_____. 1986b. "Are Cultures Living Beings: German Identity in Interaction." *Man* 21, no. 4: 587–604.

Ehrenfels, Umar Rolf. 1952. *Kadar of Cochin*. Madras: University of Madras Publications.

Evans-Pritchard, E.E. 1940. *The Nuer*. Oxford: Oxford University Press.

Fortier, Jana. 2009. "The Ethnography of South Asian Foragers." *Annual Review of Anthropology* 38: 99–114.

Fox, Richard G. 1969. "'Professional Primitives': Hunters and Gatherers of Nuclear South Asia." *Man in India* 49: 139–60.

Gardner, Peter M. 1966. "Symmetric Respect and Memorate Knowledge: The Structure and Ecology of Individualistic Culture." *Southwestern Journal of Anthropology* 22: 389–415.

_____. 1972. "The Paliyans." In *Hunters and Gatherers Today*, edited by Marco G. Bicchieri, 404–47. New York: Holt, Rinehart & Winston.

_____. 1990. "Foragers' Pursuit of Individual Autonomy." *Current Anthropology* 32, no. 5: 543–72.

_____. 2000. *Bicultural Versatility as a Frontier Adaptation among the Paliyan Foragers of South India*. Lewiston: Edwin Ellen Press.

Geertz, Clifford. 1975. *The Interpretation of Culture: Selected Essays*. London: Hutchinson.

Gray, John. 2002. *Straw Dogs: Thoughts on Humans and Animals*. London: Granta Books.

Harman, Chris. 1999. *A People's History of the World*. London: Bookmarks.

Ingold, Tim. 1996. "The Optimal Forager and the Economic Man." In *Nature and Society: Anthropological Perspectives*, edited by Philippe Descola and Gísli Pálsson, 25–44. London: Routledge.

Ingold, Tim, David Riches, and James Woodburn, eds. 1988. *Hunters and Gatherers, Vol. 1: History, Evolution and Social Change*. Oxford: Berg.

Kant, Immanuel. 2007. *Anthropology, History, and Education*. Translated by Mary Gregor et al. Cambridge: Cambridge University Press.

Kaplan, David. 2000. "The Darker Side of the 'Original Affluent Society.'" *Journal of Anthropological Research* 56, no. 3: 301–24.

Kenrick, Justin. 2005. "Equalising Processes, Processes of Discrimination and the Forest People of Central Africa." In *Property and Equality, Vol. 2: Encapsulation, Commercialisation, Discrimination*, edited by Thomas Widlok and Wolde Gossa Tadesse, 104–28. Oxford: Berghahn Books.

Kent, Susan, ed. 2002. *Ethnicity, Hunter-Gatherers and the "Other": Association or Assimilation in Africa*. Washington: Smithsonian Institute Press.

Kropotkin, Peter. 1902. *Mutual Aid: A Factor in Evolution* (1939 edition). Harmondsworth: Penguin Books.

_____. 1970. *Selected Writings on Anarchism and Revolution*. Edited by M. Miller. Cambridge, MA: MIT Press.

Leacock, Eleanor. 1983. "Primitive Communism." In *A Dictionary of Marxist Thought*, edited by T.B. Bottomore, 445–46. Oxford: Blackwell.

Leacock, Eleanor B., and Richard B. Lee, eds. 1982. *Politics and History in Band Societies.* Cambridge: Cambridge University Press.

Lee, Richard B. 1979. *The !Kung San: Men, Women, and Work in a Foraging Society.* Cambridge: Cambridge University Press.

———. 1988. "Reflections and Primitive Communism." In *Hunters and Gatherers, Vol. 1: History, Evolution and Social Change,* edited by Tim Ingold, David Riches, and James Woodburn, 252–68. Oxford: Berg.

———. 2005. "Power and Property in Twenty-First Century Foragers: A Critical Examination." In *Property and Equality, Vol. 2: Encapsulation, Commercialisation, Discrimination,* edited by Thomas Widlok and Wolde Gossa Tadesse, 16–31. Oxford: Berghahn Books.

Lee, Richard B., and Irven DeVore, eds. 1968. *Man the Hunter.* Chicago: Aldine.

Lévi-Strauss, Claude. 1977. *Structural Anthropology.* Translated by Monique Layton. Harmondsworth: Penguin Books.

Lewis, Jerome. 2005. "Whose Forest Is It Anyway? Mbendjele Yaka Pygmies, the Ndoki Forest and the Wider World." In *Property and Equality, Vol. 2: Encapsulation, Commercialisation, Discrimination,* edited by Thomas Widlok and Wolde Gossa Tadesse, 56–78. Oxford: Berghahn Books.

LiPuma, Edward. 1998. "Modernity and Forms of Personhood in Melanesia." In *Bodies and Persons: Comparative Perspectives from Africa and Melanesia,* edited by Michael Lambek and Andrew Strathern. Cambridge: Cambridge University Press.

Lukes, Steven. 1973. *Individualism.* Oxford: Blackwell.

Marx, Karl, and Frederick Engels. 1968. *Selected Works.* London: Lawrence and Wishart.

Mathur, P.R.G. 1977. *Tribal Situation in Kerala.* Trivandrum: Kerala Historical Society.

Misra, P.K. 1969. "The Jenu Kuruba." *Bulletin of the Anthropological Survey of India* 18, no. 3: 183–246.

Morgan, Lewis Henry. 1877. *Ancient Society.* Calcutta: Bharati Library.

Morris, Brian. 1978. "Are There Any Individuals in India? A Critique of Dumont's Theory of the Individual." *Eastern Anthropology* 31: 365–77.

———. 1982. *Forest Traders: A Socio-Economic Study of the Hill Pandaram.* LSE Monograph 55. London: Athlone Press.

———. 1991. *Western Conceptions of the Individual.* Oxford: Berg.

———. 1994. *Anthropology of the Self: The Individual in Cultural Perspective.* London: Pluto Press.

———. 1998. "Anthropology and Anarchism" *Anarchy* 45: 35–41.

———. 2004. *Kropotkin: The Politics of Community.* Amherst, NY: Humanity Books.

Morrison, Kathleen D. 2002. "Historicizing Adaptation, Adapting to History: Forager Traders in South and Southeast Asia." In *Forager-Traders in South and Southeast Asia,* edited by Kathleen D. Morrison and Laura L. Junker, 1–17. Cambridge: Cambridge University Press.

Murray, Henry A., and Clyde Kluckhohn. 1953. "Outline of a Conception of Personality." In *Personality in Nature, Society and Culture,* 2nd ed., edited by Clyde Kluckhohn and Henry A. Murray. New York: Knopf.

Norstrom, Christer. 2001. "Autonomy by Default versus Popular Participation: The Paliyans of South India and the Proposed Palni Hills Sanctuary." In *Identity and Gender in Hunting and Gathering Societies,* edited by Ian Keen and Takako Yamada, 27–50. Osaka: National Museum Ethnology. Senri Ethnological Studies 56.

———. 2003. *"They Call for Us: Strategies for Securing Autonomy among the Paliyan Hunter-Gatherers of the Palni Hills, South India.* Stockholm: Stockholm Studies in Social Anthropology 53.

Overing, Joanna. 1993. "The Anarchy and Collectivism of the 'Primitive Other': Marx and Sahlins in the Amazon." In *Socialism: Ideals, Ideologies and Local Practices*, edited by C.M. Hann, 43–58. London: Routledge.

Price, John A. 1975. "Sharing: The Integration of Intimate Economies." *Anthropologica* 17, no. 1: 3–27.

Rexroth, Kenneth. 1975. *Communalism: From Its Origins to the Twentieth Century*. London: Owen.

Riches, David. 2000. "The Holistic Person, or the Ideology of Egalitarianism." *Journal of the Royal Anthropological Institute* 6, no. 4: 669–85.

Rousselle, Duane, and Sürreya Evren, eds. 2011. *Post-anarchism: A Reader*. London: Pluto Press.

Rupp, Stephanie. 2011. *Forest of Belonging: Identities, Ethnicities and Stereotypes in the Congo River Basin*. Seattle: University of Washington Press.

Sahlins, Marshall. 1972. *Stone Age Economics*. London: Tavistock.

Scott, James C. 2009. *The Art of Not Being Governed: An Anarchist History of Upland Southeast Asia*. New Haven: Yale University Press.

Service, Elman. 1966. *The Hunters*. Englewood Cliffs, NJ: Prentice Hall.

Shweder, R.A. and E.J. Bourne. 1984. "Does the Concept of the Person Vary Cross-Culturally?" In *Culture Theory: Essays in Mind, Self and Emotion*, edited by R.A. Shweder and R.A. LeVine, 158–99. Cambridge: Cambridge University Press.

Solway, Jacqueline S., ed. 2006. *The Politics of Egalitarianism: Theory and Practice*. Oxford: Berghahn.

Spiro, Melford E. 1993. "Is the Western Conception of the Self 'Peculiar' within the Context of World Culture?" *Ethos* 21, no. 2: 107–53.

Strathern, Marilyn. 1988. *The Gender of the Gift*. Berkeley: University of California Press.

Van Der Walt, Lucien, and Michael Schmidt. 2009. *Black Flame: The Revolutionary Class Politics of Anarchism and Syndicalism*. Oakland: AK Press.

Weiss, Gary. 2012. *Ayn Rand Nation: The Hidden Struggle for America's Soul*. New York: St. Martin's Press.

Widlok, Thomas, and Wolde Gosse Tadesse, eds. 2005. *Property and Equality, Vol. 1: Ritualization, Sharing, Egalitarianism*. Oxford: Berghahn.

Woodburn, James. 1982. "Egalitarian Societies." *Man* 17, no. 3: 431–51.

_____. 1987. "African Hunter-Gatherer Social Organization: Is It Best Understood as a Product of Encapsulation?" In *Hunters and Gatherers, Vol. 1: History, Evolution and Social Change*, edited by Tim Ingold, David Riches, and James Woodburn, 31–64. Oxford: Berg.

Zerzan, John. 1994. *Future Primitive and Other Essays*. Brooklyn: Autonomedia.

About the Authors

Brian Morris is a professor emeritus of anthropology at Goldsmiths College, London. He received his doctorate in social anthropology at the London School of Economics and Political Science, having done his PhD field work among hunter-gatherers in Southern India. Prior to his academic career, he worked as a tea planter in Malawi, and he has carried out fieldwork there on many occasions. He has written books and articles on a wide range of topics, including ecology, botany, philosophy, history, religion, anthropology, ethnobiology, and social anarchism. After discovering anarchist thought in the mid-1960s, he remained active in various protests and political movements. His previous political books include *The Anarchist Geographer: An Introduction to the Life of Peter Kropotkin*; *Kropotkin: The Politics of Community*; *Ecology and Anarchism: Essays and Reviews on Contemporary Thought*; and *Bakunin: The Philosophy of Freedom*.

Peter Marshall is a philosopher, historian, biographer, travel writer, photographer, and poet. He is the author, among other books, of *William Godwin* (1984), *The Anarchist Writings of William Godwin* (1986, 1996), *William Blake: Visionary Anarchist* (1988, 1994, 2008), *Demanding the Impossible: A History of Anarchism* (1992, 1993, 2008, 2010), *Nature's Web: An Exploration of Ecological Thinking* (1992, 1994, 1995, 1996), *Riding the Wind: A New Philosophy for a New Era* (1998, 2000, 2009) and *Europe's Lost Civilization: Uncovering the Mysteries of the Megaliths* (2004, 2006).

Index

"Passim" (literally "scattered") indicates intermittent discussion of a topic over a cluster of pages.

ABOUT PM PRESS

PM Press was founded at the end of 2007 by a small
collection of folks with decades of publishing, media, and
organizing experience. PM Press co-conspirators have
published and distributed hundreds of books, pamphlets,
CDs, and DVDs. Members of PM have founded enduring
book fairs, spearheaded victorious tenant organizing campaigns, and worked
closely with bookstores, academic conferences, and even rock bands to deliver
political and challenging ideas to all walks of life. We're old enough to know what
we're doing and young enough to know what's at stake.

We seek to create radical and stimulating fiction and non-fiction books, pamphlets,
T-shirts, visual and audio materials to entertain, educate and inspire you. We
aim to distribute these through every available channel with every available
technology—whether that means you are seeing anarchist classics at our bookfair
stalls; reading our latest vegan cookbook at the café; downloading geeky fiction
e-books; or digging new music and timely videos from our website.

PM Press is always on the lookout for talented and skilled volunteers, artists,
activists and writers to work with. If you have a great idea for a project or can
contribute in some way, please get in touch.

PM Press
PO Box 23912
Oakland, CA 94623
www.pmpress.org

FRIENDS OF PM PRESS

These are indisputably momentous times—the financial
system is melting down globally and the Empire is
stumbling. Now more than ever there is a vital need for
radical ideas.

In the seven years since its founding—and on a mere
shoestring—PM Press has risen to the formidable challenge of publishing and
distributing knowledge and entertainment for the struggles ahead. With over
250 releases to date, we have published an impressive and stimulating array of
literature, art, music, politics, and culture. Using every available medium, we've
succeeded in connecting those hungry for ideas and information to those putting
them into practice.

Friends of PM allows you to directly help impact, amplify, and revitalize the
discourse and actions of radical writers, filmmakers, and artists. It provides us
with a stable foundation from which we can build upon our early successes and
provides a much-needed subsidy for the materials that can't necessarily pay
their own way. You can help make that happen—and receive every new title
automatically delivered to your door once a month—by joining as a Friend of PM
Press. And, we'll throw in a free T-shirt when you sign up.

Here are your options:

- **$30 a month** Get all books and pamphlets plus 50% discount on all webstore
  purchases

- **$40 a month** Get all PM Press releases (including CDs and DVDs) plus 50%
  discount on all webstore purchases

- **$100 a month** Superstar—Everything plus PM merchandise, free downloads, and
  50% discount on all webstore purchases

For those who can't afford $30 or more a month, we're introducing **Sustainer
Rates** at $15, $10 and $5. Sustainers get a free PM Press T-shirt and a 50%
discount on all purchases from our website.

Your Visa or Mastercard will be billed once a month, until you tell us to stop.
Or until our efforts succeed in bringing the revolution around. Or the financial
meltdown of Capital makes plastic redundant. Whichever comes first.

Demanding the Impossible:
A History of Anarchism

Peter Marshall

ISBN: 978-1-60486-064-1
$28.95   840 pages

Navigating the broad 'river of anarchy', from Taoism
to Situationism, from Ranters to Punk rockers, from
individualists to communists, from anarcho-syndicalists
to anarcha-feminists, *Demanding the Impossible* is an
authoritative and lively study of a widely misunderstood
subject. It explores the key anarchist concepts of society and the state, freedom
and equality, authority and power and investigates the successes and failure of
the anarchist movements throughout the world. While remaining sympathetic
to anarchism, it presents a balanced and critical account. It covers not only the
classic anarchist thinkers, such as Godwin, Proudhon, Bakunin, Kropotkin, Reclus
and Emma Goldman, but also other libertarian figures, such as Nietzsche, Camus,
Gandhi, Foucault and Chomsky. No other book on anarchism covers so much so
incisively.

In this updated edition, a new epilogue examines the most recent developments,
including 'post-anarchism' and 'anarcho-primitivism' as well as the anarchist
contribution to the peace, green and 'Global Justice' movements.

Demanding the Impossible is essential reading for anyone wishing to understand
what anarchists stand for and what they have achieved. It will also appeal to those
who want to discover how anarchism offers an inspiring and original body of ideas
and practices which is more relevant than ever in the twenty-first century.

"**Demanding the Impossible** *is the book I always recommend when asked—as I often
am—for something on the history and ideas of anarchism.*"
—Noam Chomsky

"*Attractively written and fully referenced… bound to be the standard history.*"
—Colin Ward, *Times Educational Supplement*

"*Large, labyrinthine, tentative: for me these are all adjectives of praise when applied to
works of history, and* **Demanding the Impossible** *meets all of them.*"
—George Woodcock, *Independent*

Anarchy, Geography, Modernity:
Selected Writings of
Elisée Reclus

Edited by John P. Clark and Camille Martin

ISBN: 978-1-60486-429-8
$22.95   304 pages

Anarchy, Geography, Modernity is the first comprehensive
introduction to the thought of Elisée Reclus, the great
anarchist geographer and political theorist. It shows him
to be an extraordinary figure for his age. Not only an anarchist but also a radical
feminist, anti-racist, ecologist, animal rights advocate, cultural radical, nudist, and
vegetarian. Not only a major social thinker but also a dedicated revolutionary.

The work analyzes Reclus' greatest achievement, a sweeping historical and
theoretical synthesis recounting the story of the earth and humanity as an epochal
struggle between freedom and domination. It presents his groundbreaking critique
of all forms of domination: not only capitalism, the state, and authoritarian religion,
but also patriarchy, racism, technological domination, and the domination of
nature. His crucial insights on the interrelation between personal and small-group
transformation, broader cultural change, and large-scale social organization
are explored. Reclus' ideas are presented both through detailed exposition and
analysis, and in extensive translations of key texts, most appearing in English for
the first time.

*"For far too long Elisée Reclus has stood in the shadow of Godwin, Proudhon, Bakunin,
Kropotkin, and Emma Goldman. Now John Clark has pulled Reclus forward to
stand shoulder to shoulder with Anarchism's cynosures. Reclus' light brought into
anarchism's compass not only a focus on ecology, but a struggle against both
patriarchy and racism, contributions which can now be fully appreciated thanks to
John Clark's exegesis and [his and Camille Martin's] translations of works previously
unavailable in English. No serious reader can afford to neglect this book."*
—Dana Ward, Pitzer College

*"Finally! A century after his death, the great French geographer and anarchist Elisée
Reclus has been honored by a vibrant selection of his writings expertly translated into
English."*
—Kent Mathewson, Louisiana State University

*"Maintaining an appropriately scholarly style, marked by deep background knowledge,
nuanced argument, and careful qualifications, Clark and Martin nevertheless reveal a
passionate love for their subject and adopt a stance of political engagement that they
hope does justice to Reclus' own commitments."*
—Historical Geography

Talking Anarchy

Colin Ward and David Goodway

ISBN: 978-1-60486-812-8
$14.95   176 pages

Of all political views, anarchism is the most ill-represented. For more than thirty years, in over thirty books, Colin Ward patiently explained anarchist solutions to everything from vandalism to climate change—and celebrated unofficial uses of the landscape as commons, from holiday camps to squatter communities. Ward was an anarchist journalist and editor for almost sixty years, most famously editing the journal *Anarchy*. He was also a columnist for *New Statesman*, *New Society*, *Freedom*, and *Town and Country Planning*.

In *Talking Anarchy*, Colin Ward discusses with David Goodway the ups and downs of the anarchist movement during the last century, including the many famous characters who were anarchists, or associated with the movement, including Herbert Read, Alex Comfort, Marie Louise Berneri, Paul Goodman, Noam Chomsky, and George Orwell.

"It is difficult to match the empirical strength, the lucidity of prose, and the integration of theory and practical insight in the magnificent body of work produced by the veteran anarchist Colin Ward."
—*Prospect*

"Colin Ward has never written a highly paid column for a national newspaper or been on the bestseller lists, but his fan club is distinguished, and his influence wider than he himself may know."
—*Times Literary Supplement*

A Living Spirit of Revolt:
The Infrapolitics of Anarchism

Žiga Vodovnik with an introduction
by Howard Zinn

ISBN: 978-1-60486-523-3
$15.95   232 pages

*"The great contribution of Žiga Vodovnik is that his writing
rescues anarchism from its dogma, its rigidity, its isolation
from the majority of the human race. He reveals the natural
anarchism of our everyday lives, and in doing so, enlarges*
the possibilities for a truly human society, in which our imaginations, our compassion,
can have full play."
—Howard Zinn, author of *A People's History of the United States*, from the
Introduction

At the end of the nineteenth century, the network of anarchist collectives
represented the first-ever global antisystemic movement and the very center of
revolutionary tumult. In this groundbreaking and magisterial work, Žiga Vodovnik
establishes that anarchism today is not only the most revolutionary current
but, for the first time in history, the only one left. According to the author, many
contemporary theoretical reflections on anarchism marginalize or neglect to
mention the relevance of the anarchy of everyday life. Given this myopic (mis)
conception of its essence, we are still searching for anarchism in places where the
chances of actually finding it are the smallest.

*"Like Marx's old mole, the instinct for freedom keeps burrowing, and periodically
breaks through to the light of day in novel and exciting forms. That is happening again
right now in many parts of the world, often inspired by, and revitalizing, the anarchist
tradition that is examined in Žiga Vodovnik's book. A Living Spirit of Revolt is a deeply
informed and thoughtful work, which offers us very timely and instructive lessons."*
—Noam Chomsky, MIT

*"Žiga Vodovnik has made a fresh and original contribution to our understanding of
anarchism, by unearthing its importance for the New England Transcendentalists and
their impact on radical politics in America. A Living Spirit of Revolt is interesting,
relevant and is sure to be widely read and enjoyed."*
—Uri Gordon, author of *Anarchy Alive: Anti-Authoritarian Politics from Practice to
Theory*

Anarchist Seeds beneath the Snow: Left-Libertarian Thought and British Writers from William Morris to Colin Ward

David Goodway

ISBN: 978-1-60486-221-8
$24.95   420 pages

From William Morris to Oscar Wilde to George Orwell, left-libertarian thought has long been an important but neglected part of British cultural and political history. In *Anarchist Seeds beneath the Snow*, David Goodway seeks to recover and revitalize that indigenous anarchist tradition. This book succeeds as simultaneously a cultural history of left-libertarian thought in Britain and a demonstration of the applicability of that history to current politics. Goodway argues that a recovered anarchist tradition could—and should—be a touchstone for contemporary political radicals. Moving seamlessly from Aldous Huxley and Colin Ward to the war in Iraq, this challenging volume will energize leftist movements throughout the world.

"*Anarchist Seeds beneath the Snow is an impressive achievement for its rigorous scholarship across a wide range of sources, for collating this diverse material in a cogent and systematic narrative-cum-argument, and for elucidating it with clarity and flair… It is a book that needed to be written and now deserves to be read.*"
—*Journal of William Morris Studies*

"*Goodway outlines with admirable clarity the many variations in anarchist thought. By extending outwards to left-libertarians he takes on even greater diversity.*"
—Sheila Rowbotham, *Red Pepper*

"*A splendid survey of 'left-libertarian thought' in this country, it has given me hours of delight and interest. Though it is very learned, it isn't dry. Goodway's friends in the awkward squad (especially William Blake) are both stimulating and comforting companions in today's political climate.*"
—A.N. Wilson, *Daily Telegraph*

"*The history of the British anarchist movement has been little studied or appreciated outside of the movement itself. Anarchist Seeds beneath the Snow should go a long way towards rectifying this blind spot in established labour and political history. His broad ranging erudition combined with a penetrating understanding of the subject matter has produced a fascinating, highly readable history.*"
—Joey Cain, edwardcarpenterforum.org

Anarchist Pedagogies: Collective Actions, Theories, and Critical Reflections on Education

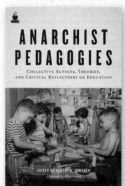

Edited by Robert H. Haworth
with an afterword by Allan Antliff

ISBN: 978-1-60486-484-7
$24.95   352 pages

Education is a challenging subject for anarchists.
Many are critical about working within a state-run
education system that is embedded in hierarchical, standardized, and authoritarian structures. Numerous individuals and collectives envision the creation of counterpublics or alternative educational sites as possible forms of resistance, while other anarchists see themselves as "saboteurs" within the public arena—believing that there is a need to contest dominant forms of power and educational practices from multiple fronts. Of course, if anarchists agree that there are no blueprints for education, the question remains, in what dynamic and creative ways can we construct nonhierarchical, anti-authoritarian, mutual, and voluntary educational spaces?

Contributors to this edited volume engage readers in important and challenging issues in the area of anarchism and education. From Francisco Ferrer's modern schools in Spain and the Work People's College in the United States, to contemporary actions in developing "free skools" in the U.K. and Canada, to direct-action education such as learning to work as a "street medic" in the protests against neoliberalism, the contributors illustrate the importance of developing complex connections between educational theories and collective actions. Anarchists, activists, and critical educators should take these educational experiences seriously as they offer invaluable examples for potential teaching and learning environments outside of authoritarian and capitalist structures. Major themes in the volume include: learning from historical anarchist experiments in education, ways that contemporary anarchists create dynamic and situated learning spaces, and finally, critically reflecting on theoretical frameworks and educational practices. Contributors include: David Gabbard, Jeffery Shantz, Isabelle Fremeaux & John Jordan, Abraham P. DeLeon, Elsa Noterman, Andre Pusey, Matthew Weinstein, Alex Khasnabish, and many others.

"Pedagogy is a central concern in anarchist writing and the free skool has played a central part in movement activism. By bringing together an important group of writers with specialist knowledge and experience, Robert Haworth's volume makes an invaluable contribution to the discussion of these topics. His exciting collection provides a guide to historical experiences and current experiments and also reflects on anarchist theory, extending our understanding and appreciation of pedagogy in anarchist practice."
—Dr. Ruth Kinna, Senior Lecturer in Politics, Loughborough University, author of Anarchism: A Beginners Guide and coeditor of Anarchism and Utopianism

Anarchism and Education: A Philosophical Perspective

Judith Suissa

ISBN: 978-1-60486-114-3
$19.95   184 pages

While there have been historical accounts of the anarchist school movement, there has been no systematic work on the philosophical underpinnings of anarchist educational ideas—until now.

Anarchism and Education offers a philosophical account of the neglected tradition of anarchist thought on education. Although few anarchist thinkers wrote systematically on education, this analysis is based largely on a reconstruction of the educational thought of anarchist thinkers gleaned from their various ethical, philosophical, and popular writings. Primarily drawing on the work of the nineteenth-century anarchist theorists such as Bakunin, Kropotkin, and Proudhon, the book also covers twentieth-century anarchist thinkers such as Noam Chomsky, Paul Goodman, Daniel Guerin, and Colin Ward.

This original work will interest philosophers of education and educationalist thinkers as well as those with a general interest in anarchism.

"This is an excellent book that deals with important issues through the lens of anarchist theories and practices of education... The book tackles a number of issues that are relevant to anybody who is trying to come to terms with the philosophy of education."
—*Higher Education Review*

The Paul Goodman Reader

Edited by Taylor Stoehr

ISBN: 978-1-60486-058-0
$28.95   500 pages

A one-man think-tank for the New Left, Paul Goodman wrote over thirty books, most of them before his decade of fame as a social critic in the Sixties. A Paul Goodman Reader that does him justice must be a compendious volume, with excerpts not only from best-sellers like *Growing Up Absurd*, but also from his landmark books on education, community planning, anarchism, psychotherapy, language theory, and poetics. Samples as well from *The Empire City*, a comic novel reviewers compared to *Don Quixote*, prize-winning short stories, and scores of poems that led America's most respected poetry reviewer, Hayden Carruth, to exclaim, "Not one dull page. It's almost unbelievable."

Goodman called himself as an old-fashioned man of letters, which meant that all these various disciplines and occasions added up to a single abiding concern for the human plight in perilous times, and for human promise and achieved grandeur, love and hope.

"It was that voice of his that seduced me—that direct, cranky, egotistical, generous American voice… Paul Goodman's voice touched everything he wrote about with intensity, interest, and his own terribly appealing sureness and awkwardness… It was his voice, that is to say, his intelligence and the poetry of his intelligence incarnated, which kept me a loyal and passionate fan."
—Susan Sontag, novelist and public intellectual

"Goodman, like all real novelists, is, at bottom, a moralist. What really interests him are the various ways in which human beings living in a modern metropolis gain, keep or lose their integrity and sense of selfhood."
—W. H. Auden, poet

"Any page by Paul Goodman will give you not only originality and brilliance but wisdom, that is, something to think about. He is our peculiar, urban, twentieth-century Thoreau, the quintessential American mind of our time."
— Hayden Carruth, poet and essayist

"No one writing now in America makes better sense of literary subjects. His ability to combine linguistic criticism, politics, a version of the nature of man, anthropology, the history of philosophy, and autobiographical testament with literary analysis, and to make a closely woven fabric of argument, seems magical."
—Robert Meredith, *The Nation*

Damned Fools in Utopia: And Other Writings on Anarchism and War Resistance

Nicolas Walter
Edited by David Goodway

ISBN: 978-1-60486-222-5
$22.95   304 pages

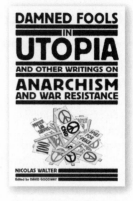

Nicolas Walter was the son of the neurologist W. Grey Walter, and both his grandfathers had known Peter Kropotkin and Edward Carpenter. However, it was the twin jolts of Suez and the Hungarian Revolution while still a student, followed by participation in the resulting New Left and nuclear disarmament movement, that led him to anarchism himself. His personal history is recounted in two autobiographical pieces in this collection as well as the editor's introduction.

During the 1960s he was a militant in the British nuclear disarmament movement—especially its direct-action wing, the Committee of 100—he was one of the Spies for Peace (who revealed the State's preparations for the governance of Britain after a nuclear war), he was close to the innovative Solidarity Group and was a participant in the homelessness agitation. Concurrently with his impressive activism he was analyzing acutely and lucidly the history, practice and theory of these intertwined movements; and it is such writings—including 'Non-violent Resistance' and 'The Spies for Peace and After'—that form the core of this book. But there are also memorable pieces on various libertarians, including the writers George Orwell, Herbert Read and Alan Sillitoe, the publisher C.W. Daniel and the maverick Guy A. Aldred. 'The Right to be Wrong' is a notable polemic against laws limiting the freedom of expression. Other than anarchism, the passion of Walter's intellectual life was the dual cause of atheism and rationalism; and the selection concludes appropriately with a fine essay on 'Anarchism and Religion' and his moving reflections, 'Facing Death'.

Nicolas Walter scorned the pomp and frequent ignorance of the powerful and detested the obfuscatory prose and intellectual limitations of academia. He himself wrote straightforwardly and always accessibly, almost exclusively for the anarchist and freethought movements. The items collected in this volume display him at his considerable best.

"[Nicolas Walter was] one of the most interesting left intellectuals of the second half of the twentieth century in Britain."
—Professor Richard Taylor, University of Cambridge

The Floodgates of Anarchy

Stuart Christie and Albert Meltzer

ISBN: 978-1-60486-105-1
$15.95   144 pages

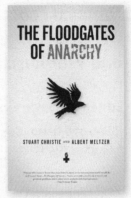

The floodgates holding back anarchy are constantly under strain. The liberal would ease the pressure by diverting some of the water; the conservative would shore up the dykes, the totalitarian would construct a stronger dam.

But is anarchy a destructive force? The absence of government may alarm the authoritarian, but is a liberated people really its own worst enemy—or is the true enemy of mankind, as the anarchists claim, the means by which he is governed? Without government the world could manage to end exploitation and war. Anarchy should not be confused with weak, divided or manifold government. As Christie and Meltzer point out, only with the total abolition of government can society develop in freedom.

"Anyone who wants to know what anarchism is about in the contemporary world would do well to start here. The Floodgates of Anarchy forces us to take a hard look at moral and political problems which other more sophisticated doctrines evade."
—The Sunday Times

"A lucid exposition of revolutionary anarchist theory."
—Peace News

"Coming from a position of uncompromising class struggle and a tradition that includes many of our exemplary anarchist militants, The Floodgates of Anarchy has a power and directness sadly missing from some contemporary anarchist writing. It is exciting to see it back in print, ready for a new generation to read."
—Barry Pateman, Associate Editor, The Emma Goldman Papers, University of California at Berkeley

Revolution and Other Writings:
A Political Reader

Gustav Landauer
edited and translated by Gabriel Kuhn

ISBN: 978-1-60486-054-2
$26.95   360 pages

"Landauer is the most important agitator of the radical
and revolutionary movement in the entire country."
This is how Gustav Landauer is described in a German
police file from 1893. Twenty-six years later, Landauer
would die at the hands of reactionary soldiers who overthrew the Bavarian Council
Republic, a three-week attempt to realize libertarian socialism amidst the turmoil
of post-World War I Germany. It was the last chapter in the life of an activist,
writer, and mystic who Paul Avrich calls "the most influential German anarchist
intellectual of the twentieth century."

This is the first comprehensive collection of Landauer writings in English. It
includes one of his major works, *Revolution*, thirty additional essays and articles,
and a selection of correspondence. The texts cover Landauer's entire political
biography, from his early anarchism of the 1890s to his philosophical reflections
at the turn of the century, the subsequent establishment of the Socialist Bund, his
tireless agitation against the war, and the final days among the revolutionaries in
Munich. Additional chapters collect Landauer's articles on radical politics in the US
and Mexico, and illustrate the scope of his writing with texts on corporate capital,
language, education, and Judaism. The book includes an extensive introduction,
commentary, and bibliographical information, compiled by the editor and translator
Gabriel Kuhn as well as a preface by Richard Day.

"If there were any justice in this world—at least as far as historical memory goes—then
Gustav Landauer would be remembered, right along with Bakunin and Kropotkin, as
one of anarchism's most brilliant and original theorists. Instead, history has abetted the
crime of his murderers, burying his work in silence. With this anthology, Gabriel Kuhn
has single-handedly redressed one of the cruelest gaps in Anglo-American anarchist
literature: the absence of almost any English translations of Landauer."
—Jesse Cohn, author of *Anarchism and the Crisis of Representation: Hermeneutics,
Aesthetics, Politics*

"Gustav Landauer was, without doubt, one of the brightest intellectual lights within
the revolutionary circles of fin de siècle Europe. In this remarkable anthology, Gabriel
Kuhn brings together an extensive and splendidly chosen collection of Landauer's
most important writings, presenting them for the first time in English translation. With
Landauer's ideas coming of age today perhaps more than ever before, Kuhn's work is a
valuable and timely piece of scholarship, and one which should be required reading for
anyone with an interest in radical social change."
—James Horrox, author of *A Living Revolution: Anarchism in the Kibbutz Movement*

Wobblies and Zapatistas: Conversations on Anarchism, Marxism and Radical History

Staughton Lynd and Andrej Grubačić

ISBN: 978-1-60486-041-2
$20.00   300 pages

Wobblies and Zapatistas offers the reader an encounter between two generations and two traditions. Andrej Grubačić is an anarchist from the Balkans. Staughton Lynd is a lifelong pacifist, influenced by Marxism. They meet in dialogue in an effort to bring together the anarchist and Marxist traditions, to discuss the writing of history by those who make it, and to remind us of the idea that "my country is the world." Encompassing a Left libertarian perspective and an emphatically activist standpoint, these conversations are meant to be read in the clubs and affinity groups of the new Movement.

The authors accompany us on a journey through modern revolutions, direct actions, anti-globalist counter summits, Freedom Schools, Zapatista cooperatives, Haymarket and Petrograd, Hanoi and Belgrade, 'intentional' communities, wildcat strikes, early Protestant communities, Native American democratic practices, the Workers' Solidarity Club of Youngstown, occupied factories, self-organized councils and soviets, the lives of forgotten revolutionaries, Quaker meetings, antiwar movements, and prison rebellions. Neglected and forgotten moments of interracial self-activity are brought to light. The book invites the attention of readers who believe that a better world, on the other side of capitalism and state bureaucracy, may indeed be possible.

"There's no doubt that we've lost much of our history. It's also very clear that those in power in this country like it that way. Here's a book that shows us why. It demonstrates not only that another world is possible, but that it already exists, has existed, and shows an endless potential to burst through the artificial walls and divisions that currently imprison us. An exquisite contribution to the literature of human freedom, and coming not a moment too soon."
—David Graeber, author of *Fragments of an Anarchist Anthropology* and *Direct Action: An Ethnography*

"I have been in regular contact with Andrej Grubačić for many years, and have been most impressed by his searching intelligence, broad knowledge, lucid judgment, and penetrating commentary on contemporary affairs and their historical roots. He is an original thinker and dedicated activist, who brings deep understanding and outstanding personal qualities to everything he does."
—Noam Chomsky

Who's Afraid of the Black Blocs? Anarchy in Action around the World

Francis Dupuis-Déri
Translated by Lazer Lederhendler

ISBN: 978-1-60486-949-1
$19.95   224 pages

Faces masked, dressed in black, and forcefully attacking the symbols of capitalism, Black Blocs have been transformed into an anti-globalization media spectacle. But the popular image of the window-smashing thug hides a complex reality. Francis Dupuis-Déri outlines the origin of this international phenomenon, its dynamics, and its goals, arguing that the use of violence always takes place in an ethical and strategic context.

Translated into English for the first time and completely revised and updated to include the most recent Black Bloc actions at protests in Greece, Germany, Canada, and England, and the Bloc's role in the Occupy movement and the Quebec student strike, *Who's Afraid of the Black Blocs?* lays out a comprehensive view of the Black Bloc tactic and locates it within the anarchist tradition of direct action.

"A level-headed, carefully researched inquiry into a subject that reduces most pundits to foaming at the mouth."
—CrimethInc. Writers' Bloc

"Francis Dupuis-Déri's discussion of Black Blocs is intimately well-informed, truly international in scope, and up-to-the-minute. He treats the complex issues surrounding the tactic with an admirable balance of sympathy and sobriety. This book is the ideal antidote to the misinformation spread by the establishment, its defenders, and its false critics."
—Uri Gordon, author of *Anarchy Alive!*

"Wearing black to mask their identities, the Black Bloc fights injustice globally. Although little is known about these modern Zorros, this book critically reveals their origins and prospects. I heartily recommend it."
—George Katsiaficas, author of *The Subversion of Politics*

"The richness, imaginativeness, and sheer learning of Francis Dupuis-Déri's work is stimulating and impressive. The whole book turns on a fascinating blend of the rigorously analytical and the generously imaginative. It was high time that it should be translated into English, as this well-established anarchist classic will both delight and inform."
—Andrej Grubačić, Professor of Anthropology and Social Change, California Institute of Integral Studies, and coauthor of *Wobblies & Zapatistas*

William Morris: Romantic to Revolutionary

E.P. Thompson
with a foreword by Peter Linebaugh

ISBN: 978-1-60486-243-0
$32.95   880 pages

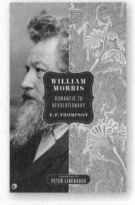

William Morris—the great 19th century craftsman,
architect, designer, poet and writer—remains a
monumental figure whose influence resonates
powerfully today. As an intellectual (and author of
the seminal utopian *News from Nowhere*), his concern with artistic and human
values led him to cross what he called the 'river of fire' and become a committed
socialist—committed not to some theoretical formula but to the day by day
struggle of working women and men in Britain and to the evolution of his ideas
about art, about work and about how life should be lived.

Many of his ideas accorded none too well with the reforming tendencies dominant
in the Labour movement, nor with those of 'orthodox' Marxism, which has looked
elsewhere for inspiration. Both sides have been inclined to venerate Morris rather
than to pay attention to what he said.

Originally written less than a decade before his groundbreaking *The Making of the
English Working Class*, E.P. Thompson brought to this biography his now trademark
historical mastery, passion, wit, and essential sympathy. It remains unsurpassed as
the definitive work on this remarkable figure, by the major British historian of the
20th century.

"*Two impressive figures, William Morris as subject and E. P. Thompson as author, are
conjoined in this immense biographical-historical-critical study, and both of them have
gained in stature since the first edition of the book was published… The book that was
ignored in 1955 has meanwhile become something of an underground classic—almost
impossible to locate in second-hand bookstores, pored over in libraries, required
reading for anyone interested in Morris and, increasingly, for anyone interested in
one of the most important of contemporary British historians… Thompson has the
distinguishing characteristic of a great historian: he has transformed the nature of the
past, it will never look the same again; and whoever works in the area of his concerns in
the future must come to terms with what Thompson has written. So too with his study
of William Morris.*"
—Peter Stansky, *The New York Times Book Review*

"*An absorbing biographical study… A glittering quarry of marvelous quotes from
Morris and others, many taken from heretofore inaccessible or unpublished sources.*"
—Walter Arnold, *Saturday Review*